建筑立场系列丛书 No.43

休闲小筑
Vacation Stay

中文版
(韩语版第359期)

韩国C3出版公社 | 编

时真妹 曹硕 楚立峰 谭文君 周一 栾一斐 于风军 张琳娜 | 译

大连理工大学出版社

建筑立场系列丛书 No.43

陆地赛艇

004 陆地赛艇 _ *Paula Melâneo*

016 克拉克公园的WMS船屋 _ Studio Gang Architects

016 波西尼奥高性能赛艇中心 _ spacialAR-TE

休闲小筑

030 休闲住宅 _ *Aldo Vanini*

034 Timmelsjoch体验馆 _ Werner Tscholl Architekt

058 山脊咖啡馆 _ Harris Butt Architecture

066 帕德尔内的住宅 _ Carlos Quintáns Eiras

078 福古岛的旅馆 _ Saunders Architecture

086 丽阿海滩度假住宅 _ Mold Architects

098 安布兰海滩俱乐部 _ Erginoğlu & Çalışlar Architects

106 维拉多姆斯儿童夏令营 _ OAB – Office of Architecture in Barcelona

114 圣乔治侦察空间 _ Mutar Arquitectos

124 朝圣旅馆 _ Sergio Rojo

住宅与社区

132 提高多孔性：重新思考住宅与社区之间的交界面 _ *Nelson Mota*

136 Dragon Court村庄 _ Eureka

152 阿尔卡比德希的社会建筑群 _ Guedes Cruz Arquitectos

166 Ulus Savoy住宅 _ Emre Arolat Architects

178 Version Rubis住宅 _ Jean-Paul Viguier et Associes

186 建筑师索引

Rowing in Dry

004 *Rowing in Dry _ Paula Melâneo*

016 WMS Boathouse at Clark Park _ Studio Gang Architects

016 Pocinho High Performance Rowing Center _ spacialAR-TE

Vacation Stay

030 *Free-time Dwellings _ Aldo Vanini*

034 The Timmelsjoch Experience _ Werner Tscholl Architekt

058 Knoll Ridge Cafe _ Harris Butt Architecture

066 House in Paderne _ Carlos Quintáns Eiras

078 Fogo Island Inn _ Saunders Architecture

086 Lia Beach Vacation Residence _ Mold Architects

098 Amburan Beach Club _ Erginoğlu & Çalışlar Architects

106 Viladoms Children's Summer Camps _ OAB – Office of Architecture in Barcelona

114 Saint George Scout Space _ Mutar Arquitectos

124 Pilgrim Hostel _ Sergio Rojo

Dwelling and Community

132 *Enhancing Porosity: Rethinking the Interface Between the Dwelling and the Community*
   _ Nelson Mota

136 Dragon Court Village _ Eureka

152 Social Complex in Alcabideche _ Guedes Cruz Arquitectos

166 Ulus Savoy Housing _ Emre Arolat Architects

178 Version Rubis Housing _ Jean-Paul Viguier et Associes

186 Index

陆地赛艇
Rowing in Dry

根据法老墓碑上的文字记载,赛艇被认为是一项源于古埃及的古老运动。它在1900年被纳入比赛,是(近代时期)最古老的奥林匹克运动项目之一。

尽管如此,赛艇却从没有像被高额的经济和商业利益所驱动的足球运动那样,被视为标杆性运动项目。因此,相关的配套基础设施大部分存放于多用途的仓库中,那里既用于储藏,也用于室内相关活动,以及小棚屋中,而在建的小棚屋也作为支持开展室外活动的一种快速解决方案。但是,有一处专属空间,设计智能、简约,能够使个人或团队的训练达到最佳效果。房屋内设有像测功机、荡桨池及特种训练中需要用到的其他设施。

作为一项卓越的户外运动,赛艇运动理想的训练场地是大面积的水面,像是湖泊、江河或大海,而有时是在一些最迷人、最令人振奋的景区里。可持续发展的方法能够使每一处场所物尽其用。

在后面关于近代体育设施的两个项目全部都用来介绍了赛艇运动设施,文中涉及的两个项目都是出于当地政府的意愿而促成的。场所之一位于葡萄牙东北部,该工程需要克服地势障碍,贴合世界文化遗址的巨大表面来建造。另一个项目则需要应付一个复杂的环境,地处芝加哥城一处疏于管理的地区,正挨着一条长期受污染的河流。该场地是一个大规模河滨振兴计划中的一部分,目标是让河边地区转向公众开放,并开辟出一处新的休闲区。

波西尼奥高性能赛艇中心由葡萄牙的spacialAR-TE建筑事务所设计,坐落于多罗河地区,自2001年起被列为联合国教科文组织世界文化遗产地。这处美妙的情境是著名的波特酒产区——多罗河谷一处用来栽种葡萄的梯田地势景观。建筑物靠近波西尼奥水坝,建于20世纪80年代,被认为是赛艇运动的最佳地点之一,国内及国际奥林匹克运动员都强烈要求在这里比赛。

该建筑的形式复杂,随地势呈波浪状起伏的外形是对当地景观的重新解读。为把这个8000m²的项目对周围环境的影响降到最低,设计团队找到了这样的解决方法——沿着山谷斜坡,将建筑的一部分埋于地下。建筑的公用部分,一组白色的体量是专门为社交联谊活动设计的,位于场地的上层,在此能够俯瞰河谷雄伟震撼的景色。训练设施设置在较低层,临近河边的另一座白色建筑里,有助于运动员将精力集中在高性能的赛艇训练活动中。在两个场馆之间,宿舍通过几级台阶的连接直线分布,宿舍封闭,向内,简约得像僧舍一样。

梯田区域使得中心的住宅设施在未来有条件扩大范围,欢迎更多的运动员来这里训练。

Figuring in pharaohs' funerary monuments, rowing is considered an ancient sport, with its origin in the ancient Egypt. It is one of the oldest(modern era) Olympic sports, which was integrated into competitions in 1900.

Although rowing has never been seen as a heading sport as, for example, football, where the economic and commercial benefits that surround it are very high. Thus, supporting infrastructures mostly consisted in adapting warehouses, for storage and indoor-related activities, and in constructing small huts as a fast solution to support the outdoor activities. But a dedicated space, with an intelligent and simple design, can optimize uses for individual or team workout, housing specific machinery such as ergometer machines, rowing water tanks and other useful features for specialized training.

As an outdoor sport by excellence, rowing is, ideally, to be practiced in large water surfaces, such as lakes, rivers or sea, and therefore in some of the most fantastic and inspiring landscapes. A sustainable approach can make the best of what each place has to offer.

Both projects presented in the next pages refer to recent sports facilities, entirely dedicated to rowing activities, that were made possible by local political will. One, located in the northeast of Portugal, has the hard task of conciliating a large surface program with a World Heritage Site. The other project deals with a complex context consisting on a neglected area in the city of Chicago, just beside a long-polluted river. This site is part of a larger revitalization program, with such a positive goal for turning the riverfront accessible to the public and implementing a new recreational area.

Pocinho High Performance Rowing Center, designed by the Portuguese practice of spacialAR-TE, is located in the Douro River area, classified as a UNESCO World Heritage Site since 2001. This wonderful scenario is the area where the famous Port wine is produced, within a valley landscape shaped with terraces for the culture of the vine. The building is located near the Pocinho Dam, constructed in the 1980s, which made the place being considered as one of the best spots for rowing, extremely requested by national and international Olympic athletes.

The construction has a complex form, as a reinterpretation of the landscape, where the general waving shape accompanies the topography. Burying part of the construction, along the valley slope, was the solution found by the design team to minimize the impact that could result from containing the 8,000m² program. The building's common parts, a set of white volumes designed for social gathering, are located on the upper level of the plot allowing great and inspiring views over the valley. The training facilities, contained in another white construction set, are at the lower level, closer to the river, for a focused mindset on the rowing – a high performance activity. In between those two areas, linearly distributed by several steps, are the dormitories. Just like monk's cells, those are simple spaces with closed and introspective characteristics. This terracing area allows the future expansion of the housing facilities, so the center can welcome more athletes for training.

克拉克公园的WMS船屋_WMS Boathouse at Clark Park/Studio Gang Architects
波西尼奥高性能赛艇中心_Pocinho High Performance Rowing Center/spacialAR-TE
陆地赛艇_Rowing in Dry/Paula Melâneo

波西尼奥赛艇中心也是一项能量集中型的设计作品。多罗河谷地区夏季炎热干燥,冬季寒冷潮湿,因此设计师主要考虑了要让被动能量系统随季节变换,从而得到最优化使用。窗户根据规划进行敞开,引入太阳热能或提供凉爽荫蔽。覆盖在宿舍上的绿色屋顶起到了有效的保温作用,天窗引入阳光,提供采光和采暖的同时还能保持自然通风。

为满足"方便所有人自由活动和进出"的需要,入口大厅设有平级的入口,方便残障运动员进出和活动。

尽管占据了大面积区域,综合设施在某种程度上仍然像一座寺院一样,相对封闭。周围环境优美至极,让人忍不住想要出去休闲放松一下,然而这里不是一处度假营地,而是一个工作场所。肃净的环境带有分散设置的出入口,在这里运动员能够为比赛做好准备工作,并将注意力高度集中在提高自身技能上。

相反,克拉克公园的WMS船屋由北美的Gang建筑师工作室设计,是一座更加向公众与游客开放的建筑。它竭力吸引芝加哥的市民、学生和游客到此欣赏河畔美景以及新开发的各种相关的精彩活动。两种不同的体量形成一幅动态的图景,紧依河畔,构成了整座建筑。第一眼看上去,建筑物会使人们联想到工厂的特征,带有尖角的屋顶和大型朝向阳光开放的洞口。然后人们就会明白,波浪形的屋顶是以运动员划船时有节奏韵律的动作为原型,受摄影师埃德沃德·迈布里奇著名的定格拍摄手法启发而设计的。

两座建筑——单层的船艇储藏间和两层的训练场馆及辅助区域,都大面积地镶嵌了玻璃,这样,储藏间上方的天窗就能为室内带来自然采光和采暖,"在冬日,楼板变得暖洋洋,在夏季,保持了室内通风,将全年能耗降到最低",Gang工作室这样解释。尽管如此,考虑到主要用于人们住宿和活动(而不是储藏船艇),这座两层的建筑物还是配备了加热和制冷设备。一楼立面处的休息室设有一个开放的巨大的木质阳台,搭建在河上方,是观看赛艇活动极好的位置。

灰色楼群、石板瓦和镀锌面板的使用使船屋的立面看起来十分牢固,且呈深色。而室内天花板和墙面则在普通的混凝土底板上覆以胶合板,给人如船舱一样的舒适感受。内外氛围形成强烈反差。

WMS船屋如今作为河边地区的一个新地标,属于河滨地区一个更广阔的发展计划的一部分。外部景观的翻修工作,比如建筑之间的庭院和远及浮动的始发码头的河滨振兴计划,意在弥补建筑物之不足,将此区域改造为一处令人愉悦、极具魅力并且热闹非凡的公共休闲场所。Gang建筑师工作室目前正忙于该地区另一个船屋的设计工作,该项目预计将在2015年竣工。

Pocinho Center is also an energy concerned design, mainly regarding passive energy systems for optimizing consumptions along the seasonal changes, between Douro's hot dry summer and a cold humid winter. Windows are strategically opened, allowing the solar heat or providing a refreshing shadow. The green-roof over the dormitories works as efficient insulation, with skylights for solar lighting and heat and also for natural ventilation.

Respecting the "Mobility and Accessibility for All" requisites, athletes with physical disabilities have equal access to the rooms at the access hall level, as they are easily adaptable.

Even if occupying a large area, the complex is somehow closed to the exterior, almost as a monastery. The environment is fantastic, inviting for leisure, yet it is not a holiday campus but a workspace. Sober, with discrete openings, here athletes can prepare themselves to compete and be highly concentrated on their accomplishment.

On the contrary, the WMS Boathouse at Clark Park, designed by the north American Studio Gang Architects, is a structure more open to the public and visitors. It tries to seduce Chicago citizens, students and tourists to enjoy the river and the newly created activities related to it. Two different volumes, with a dynamic drawing, just by the river, compose the building. At a first glance the construction reminds us of a factory feature, with the sharp angle roof and the upper generous openings to the light. But then we understand that the roof is waving, taking its shape from the rhythmic cadence of the movement of an athlete's rowing, as being inspired by Eadweard Muybridge's famous stop-motion photographs.

Both volumes, the single-storey boat storage and the two-storey volume for training and supporting spaces, have generous glazed areas. The upper clerestory of the storage volume brings natural light and heat to the interior. "It warms the floor slab of the structure in winter and ventilates in summer to minimize energy use throughout the year", explains Studio Gang. Though, the two-storey building is equipped with heating and cooling systems, considering that it is mainly used to house people and activities (and not for boats' storage). A retreat of the facade on the first floor opens a large wood balcony over the river, resulting in an excellent opportunity of watching the rowers' activity.

The boathouse's exterior facades have a strong and dark image, where grey blocks, slate shingles and zinc panels are used, contrasting with the lighter and cosy interior atmosphere, where ceilings and walls are covered with plywood panels, over a general concrete flooring.

The WMS Boathouse functions now as new landmark in the riverfront, part of a wider development plan for the river area. The exterior landscape's re-fitments, such as the court between the buildings and the waterfront plan over the floating launch dock, complement architecture and transform the area in a pleasant, inviting and lively leisure public place. Studio Gang Architects is engaged in the design of another boathouse for the area, previewed to be completed in 2015. Paula Melâneo

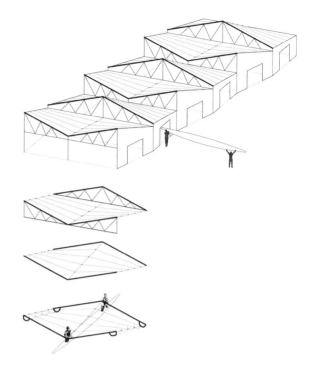

克拉克公园的WMS船屋

Studio Gang Architects

当芝加哥城正致力于将长期受污染的芝加哥河改造成为下一个娱乐新前沿时,克拉克公园的WMS船屋为其树立了重要的里程碑。船屋的建筑风格被刻意打造成了有着诗般的韵律和划船动作的视觉效果。通过在河边地区创造一条新建的、直达的公众通道,来揭晓一个围绕着河流生态和娱乐复兴展开的更大举措。通过这种途径,建筑结构中的动作捕捉所表现出来的不仅是桨手的进步,同时也表达了促进公众利益,提高生活质量和环境公正性之类的重要事项。

船屋的结构是双向的——面向河流,同时彼此相对,这样从任何视角看上去都能获得视觉上的愉悦之感。建筑物环抱着一个私密的迎宾庭院,与一段宽阔的倾斜的阶地(及远处的一个浮动的始发码头)连接,被透水混凝土和沥青以及种植了本地物种的开阔绿地包围,有利于减少水土流失及地下水补给。

船屋目前是芝加哥赛艇基金会(CRF)的大本营。它提供了一处有趣的、刺激的空间,让全城的人们,尤其是年轻人,都能够参与到运动中来,让人们在发展重要的生活技能之余还能直接亲近自然,为芝加哥都市社区增添了活力。该建筑全年提供多种室内、室外活动,包括在池塘和河上学习划船、团队划船、测功计训练、受划船运动启发而开发的瑜伽课,以及针对残障人士量身定制的特殊训练课程。芝加哥赛艇基金会与芝加哥公园区合作,担当着社区的管理者,在每年夏天主办青年夏令营,组织社区活动,推进来自29所高中及36所初中的学生的活动项目。

船屋屋顶形态的设计旨在通过建筑学的手段来诠释定格了的划船动作,给人以视觉趣味性的同时也提供了空间和环境的优势,让船屋能够适应芝加哥独特的季节变迁。起伏的桁架结构是倒"V"和"M"形状的交替,南侧的光线自上层的天窗漏下,起到了有节律的调节作用。天窗玻璃温暖了冬日的地板,并能在夏季保持通风,将全年能耗降到最低。

2101m²的综合设施包含两层可机械加热和制冷的训练场馆、一层的船艇存储设施,以及一个43m长的浮动始发码头,并且带有一个特别设计的12m宽的舷梯。主建筑物里设有室内荡桨池、测功机、共享空间和一间芝加哥公园区办公室,饰以河畔的华丽美景作为开阔的甲板。船艇储存区包括皮艇和独木舟的供应商专属区、一间办公室,还有放置赛艇及辅助设备的净跨仓库。

建筑外层覆以镀锌面板和晕石南沼泽出产的板岩,这种板岩是一种特别坚固且耐用的材料,吸水率低,能够抵抗得住冻融循环并且对环境无害。晕石南沼泽出产的板岩也被应用在可调的动态屋顶上。而在建筑内部,加拿大花旗松板材天花板和预制的木质建筑材料都会使人联想到船体结构。全部的地板都是具有高度耐受性和持续性的洋槐木材质的。

整个项目自始至终,从方案设计直至整个建设阶段中,土木工程师、滨水区工程师和景观建筑师都在团队中起到了不可替代的作用。雨水管理技术被应用到项目中,以协助应对场地的气候特性。下雨时,雨水在园林和透水混凝土及沥青下的石洞中被就地保留并储存下来,园区的绝大部分硬质景观都是由这两种材料制成的。过滤的雨水以0.055m³/s的速度缓慢回流到河里。施工期间,防水土工织物被堆放在现有土壤的周围,以防止凝固和侵蚀。项目团队与社区合作来还原该地区的草地、原生植物和树木群落。现有的生态环境通过本地植物和河边的可透水景观缓冲带得以维持和增强。

WMS Boathouse at Clark Park

As the City of Chicago works to transform the long-polluted Chicago River into its next recreational frontier, the WMS Boathouse at Clark Park helps catalyze necessary momentum. The boathouse

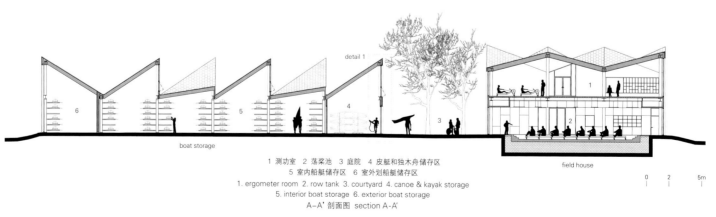

1 测功室 2 荡桨池 3 庭院 4 皮艇和独木舟储存区
5 室内船艇储存区 6 室外划船艇储存区
1. ergometer room 2. row tank 3. courtyard 4. canoe & kayak storage
5. interior boat storage 6. exterior boat storage
A-A' 剖面图 section A-A'

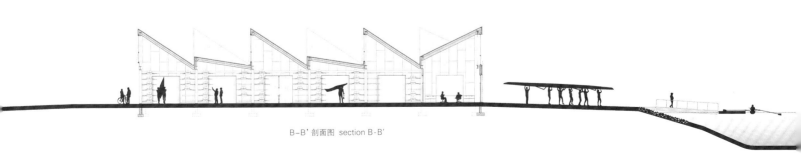

B-B'剖面图 section B-B'

项目名称：WMS Boathouse at Clark Park
地点：3400 North Rockwell Ave, Chicago, IL, USA
建筑师：Studio Gang Architects
发起人和设计主管：Jeanne Gang / 主管负责人：Mark Schendel / 项目建筑师：William Emmick
设计团队：John Castro, Juan de la Mora, Jay Hoffman, Wei-Ju Lai, Angela Peckham, Christopher Vant Hoff, Michan Walker, Todd Zima
结构工程师：Matrix Engineering Corporation / 土木工程师：Spaceco, Inc. / 河流土木工程师：AECOM
MEP和照明工程师：dbHMS / 景观建筑师：Terry Guen Design Associates, Inc.
总承包商：Schaefges Brothers, Inc. / 屋顶承包商：M Cannon Roofing
板岩和锌材料承包商：Mortensen Roofing Co., Inc. / 胶合板承包商：Wendell Builders
甲方：Chicago Park District / 面积：2,100m² 造价：USD 8.8 million / 竣工时间：2013
摄影师：©Steve Hall / Hedrich Blessing(courtesy of the architect)

is intended to visually capture the poetic rhythm and motion of rowing. By creating a new, direct public access point along the riverfront, it also reveals the larger movement toward an ecological and recreational revival of the river. In this way, the motion captured in the structure speaks not only to rowers' progress but also to advance important issues of public interest, quality of life, and environmental justice.

Visually pleasing from any viewpoint, the boathouse's structures are dually oriented – toward the river and toward each other. The buildings embrace an intimate, welcoming courtyard that links to an expansive, sloping terrace (and a floating launch dock beyond), delineated with pervious concrete and asphalt and wide open green spaces planted with native species to reduce runoff and allow for groundwater recharge.

Currently home to the Chicago Rowing Foundation(CRF), the boathouse supports the vitality of Chicago's urban communities by providing a fun and stimulating space where people from all over the city, especially young people, can participate in sport and develop important life skills while also directly connecting to nature. The building accommodates a wide range of indoor and outdoor activities year-round, including learn-to-row sessions both in tanks and on the river, team rowing, ergometer training, rowing-inspired yoga classes, and lessons tailored to individuals with disabilities. In partnership with the Chicago Park District, the CRF acts as a community steward by hosting youth campers each summer, organizing community events, and facilitating programming for students from twenty-nine high schools and thirty-six middle schools.

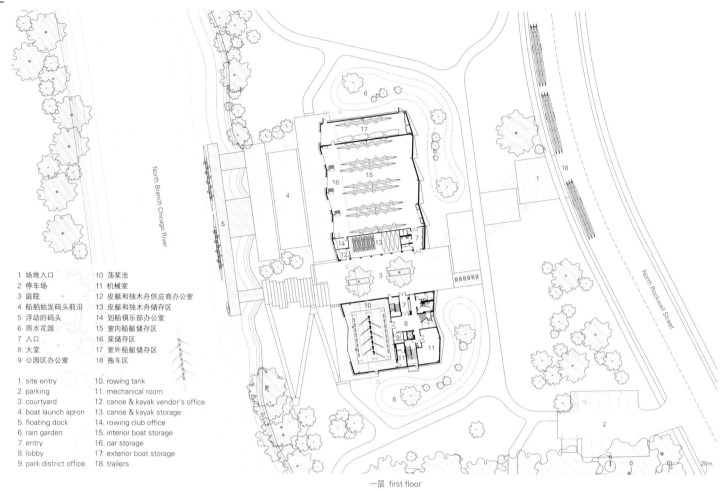

1 场地入口	10 荡桨池	1. site entry	10. rowing tank
2 停车场	11 机械室	2. parking	11. mechanical room
3 庭院	12 皮艇和独木舟供应商办公室	3. courtyard	12. canoe & kayak vendor's office
4 船舶始发码头前沿	13 皮艇和独木舟储存区	4. boat launch apron	13. canoe & kayak storage
5 浮动的码头	14 划船俱乐部办公室	5. floating dock	14. rowing club office
6 雨水花园	15 室内船艇储存区	6. rain garden	15. interior boat storage
7 入口	16 桨储存区	7. entry	16. oar storage
8 大堂	17 室外船艇储存区	8. lobby	17. exterior boat storage
9 公园区办公室	18 拖车区	9. park district office	18. trailers

一层 first floor

The design of the boathouse translates the time-lapse motion of rowing into an architectural roof form, providing visual interest while also offering spatial and environmental advantages that allow it to adapt to Chicago's distinctive seasonal changes. Alternating between an inverted "V" and an "M," the roof achieves a rhythmic modulation that lets in southern light through the building's upper clerestory. The clerestory glazing warms the floor slab of the structure in winter and ventilates in summer to minimize energy use throughout the year. The 22,620-square-foot complex consists of a two-story mechanically heated and cooled training center; one-story boat storage facility; and a 140-foot-long floating launch dock, custom-engineered with a forty-foot-wide gangway. The main building houses indoor rowing tanks, ergometer machines, communal space, and an office for the Chicago Park District, and features a spacious deck with gorgeous views of the river. Boat storage includes space for kayak and canoe vendors, an office, and clear span storage space for rowing shells and supporting equipment.

The exterior is clad in zinc panels and Heathermoor slate, an exceptionally strong and durable material with a low absorption rate, resistance to freeze-thaw cycles, and no detrimental effects on the environment. Heathermoor slate is also used to shingle the dynamic, modulating roof. In the interior, the sloping Douglas Fir plywood ceilings and customized millwork suggest the hull of a boat. Highly durable and highly sustainable black locust wood was used for the flooring throughout.

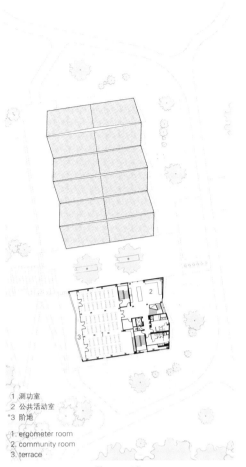

1 测功室
2 公共活动室
*3 阶地

1. ergometer room
2. community room
3. terrace

二层 second floor

Civil engineers, waterfront engineers, and landscape architects were integral members of the team throughout the entirety of the project, from schematic design through construction. Stormwater management techniques were selected collaboratively and tailored to the specifics of the site. Rainwater is retained and stored on-site in rain gardens and stone voids below the permeable concrete and asphalt that form the vast majority of the hardscape. Filtered rainwater is slowly returned to the river at a rate of 1.95 cubic feet per second. During construction, silt fabric was placed around existing soil stock piles to prevent compaction and erosion. The team worked with the community to restore a mix of grass and native plants and trees to the site. Existing habitats were maintained and strengthened with a native plant and permeable landscape buffer at the river's edge.

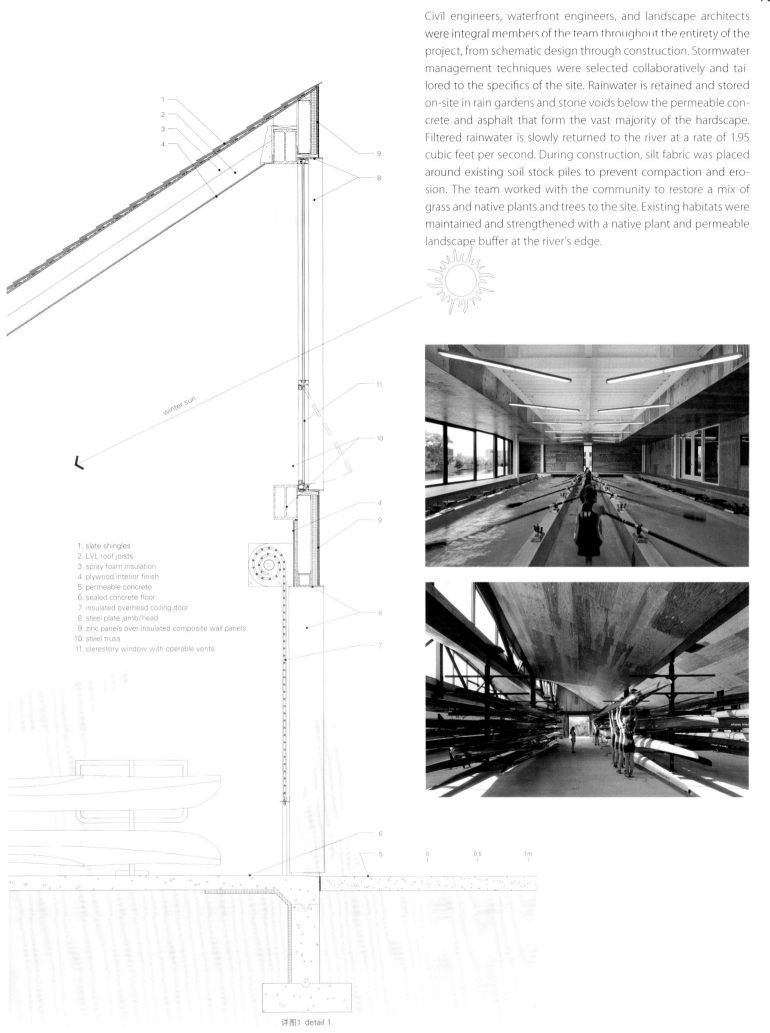

1. slate shingles
2. LVL roof joists
3. spray foam insulation
4. plywood interior finish
5. permeable concrete
6. sealed concrete floor
7. insulated overhead coiling door
8. steel plate jamb/head
9. zinc panels over insulated composite wall panels
10. steel truss
11. clerestory window with operable vents

详图1 detail 1

项目名称：Pocinho High Performance Rowing Center
地点：Pocinho, Foz Coa, Portugal
建筑师：Álvaro Fernandes Andrade
合作伙伴：Ana Rute Costa, André Bevilacqua, Marcelo Altieri, Daniel Geada, Nilton Marques, Luís Romero, Paula Cicuto, Daniela Teixeira
协调设计师：Paula Teles
合作设计师：Adelino Ribeiro, Adriana Sá, Daniel Geada, Jorge Gorito, Pedro Ribeiro da Silva
结构顾问：Machado dos Santos
电气顾问：Manuel Costa Simões / 机械顾问：José Rocha
景观建筑师：Álvaro Fernandes Andrade, Machado dos Santos
承包商：Manuel Vieira & Irmãos, Lda.
甲方：Câmara Municipal de Vila Nova de Foz Côa
面积：8,000m²
施工时间：2007~under construction
摄影师：©FG+SG Architectural Photography(except as noted)

波西尼奥高性能赛艇中心

spacialAR-TE

北立面 north elevation

波西尼奥高性能赛艇中心项目的指导性原则和策略是将其打造成为一座高度密集并且集多种用途为一体的训练场馆,包括先已存在的特定"场所"的特殊性与一致性;对于新建项目的特性和需求;同时还有对于建筑艺术的需求和期待。

首先,关于这个场所的特殊性,从广义上讲,主要关乎多罗河谷作为联合国教科文组织世界文化遗产这一文化层面的事实,在经过了人工的干预和景观改造后,更能够明确地表达出对当地祖先的敬意。

至于第二点,一个新建项目(高性能运动中心)需要获得建筑界的关注,在很多方面我们可以说,就像歌手Sting很久之前在歌曲中唱到的一样,"没人能把想法强加于总统,历史上没有这样的先例"。

但考虑到设计的需要(就像建筑风格也不是意愿与创新的有意识行为),建筑师反过来也必须在"预先设定的"必要条件之内(比如要确保"方便所有人自由活动和进出"的需要和时时存在的"可持续发展"的基本价值观念)完成,并且在整个设计过程中将这些要求具体化。在这些原则中我们最后才会考虑这一项大型工程的占地问题(总面积8000m²/共84个房间/约入住130位住户),未来计划在随之而来的居住面积的扩展中,还要进行扩建(达到总面积11 500m²/拥有170个房间/约入住225位住户)。但这已然不会涉及加大项目征地,也不会影响周边的景观。结果,在复杂的相互影响下,建筑师决定将项目分为三个基本分区(社交区、居住区和训练区),并与多罗河谷地区的世俗建筑的两种元素融合,进行了重新解读:这两种元素分别为无处不在的梯田地貌,喻意一种显著倾斜的山谷中的生存循环模式(这里的"生存"指的是"从土地中获取食物"),以及建在那些梯田地貌中间的大规模白色建筑群。尤其是那些大型酿酒厂,体积和结构都很复杂(往往是由于建筑在很长一段时期为了耕地种植的需要而交替变迁造成的)。

建筑师在梯田和建筑之间创造了一种生硬的串连,这种连接将梯田、陡斜的坡道、墙壁与墙壁之间的楼梯变相地撕裂开来,以满足规划的需要。

以上所述是波尔图大学建筑学院对于建筑学历史的典型解读,其本身并不是终点,而是与其他设计难题一起作为被带到了制图板/计算机上的又一个要素。

训练中心项目令人着迷并且激动人心,是一项对建筑形式的挑战,同时也是一个将多重"新"主题的特异性相融合的过程方面的挑战,比如"可达性"和"可持续发展",这是建筑师真正追求的,是建筑师真正想要诠释的,任何修饰的词语都只会削减它。它不是"环保的"或"绿色的",也不是"可达的"或"可持续发展的"。建筑,真正的建筑,其本身即是全部,甚至还要多得多。

Pocinho High Performance Rowing Center

The guiding principles and strategies of the project for the Pocinho Center for High Performance Rowing play their part in a dense and inextricable mixture that includes the peculiarities and identity of a pre-existing, specific "place"; the characteristics and demands of a very recent program; and the needs and wants of the architectural act.

The first one, about the specificity of the site, taken in a broad sense, as a "cultural being", concerns mainly the Douro River Valley as a UNESCO World Heritage Site, and the correspondent ancestral expression of men's intervention and transformation of the landscape.

Regarding the second, being a new program demanding architectural attention (the High Performance Sports Centers), in many ways we can say, as Sting's song a "few" years ago, "there's no

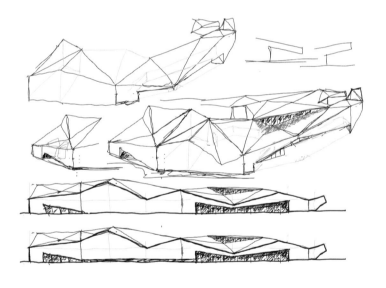

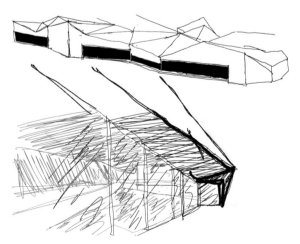
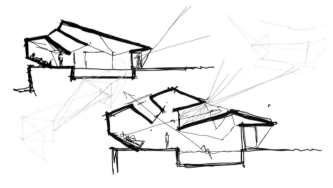

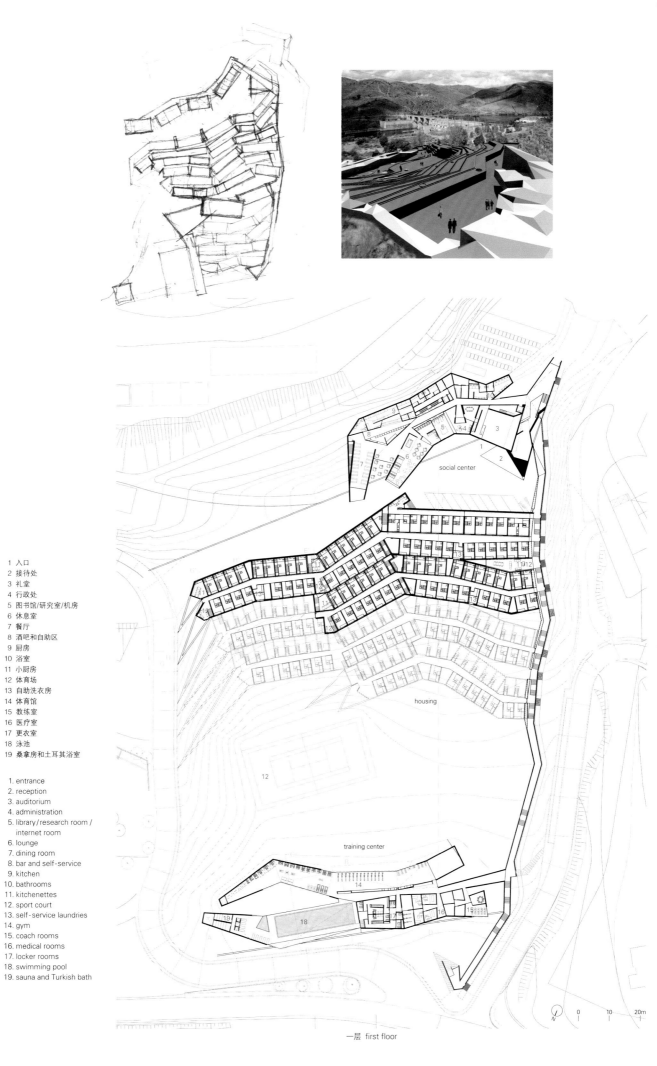

1 入口
2 接待处
3 礼堂
4 行政处
5 图书馆/研究室/机房
6 休息室
7 餐厅
8 酒吧和自助区
9 厨房
10 浴室
11 小厨房
12 体育场
13 自助洗衣房
14 体育馆
15 教练室
16 医疗室
17 更衣室
18 泳池
19 桑拿房和土耳其浴室

1. entrance
2. reception
3. auditorium
4. administration
5. library/research room/internet room
6. lounge
7. dining room
8. bar and self-service
9. kitchen
10. bathrooms
11. kitchenettes
12. sport court
13. self-service laundries
14. gym
15. coach rooms
16. medical rooms
17. locker rooms
18. swimming pool
19. sauna and Turkish bath

一层 first floor

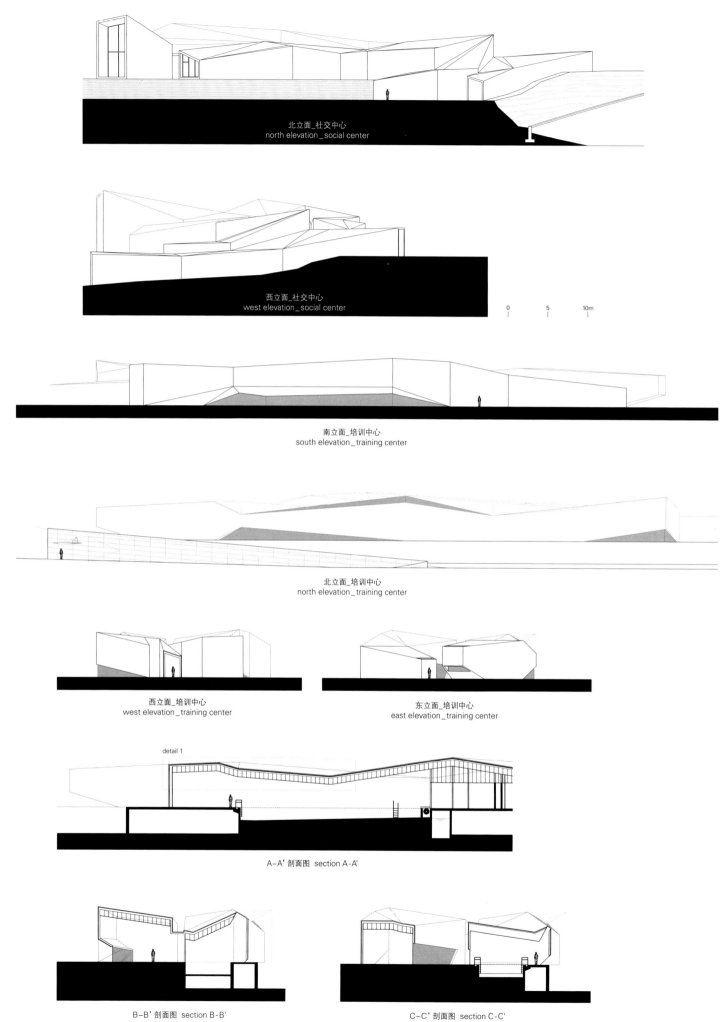

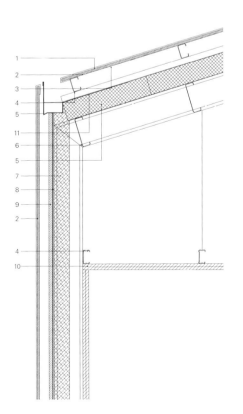

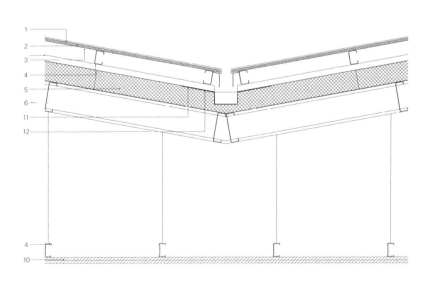

1. constructive finish
2. aquapanel
3. tyvek fabric
4. fixing metal profiles
5. 100mm sandwich panel of galvanized insulation sheet
6. LSF structure
7. isofloc 80mm, projected
8. pavaplan 8mm
9. isoroof natur 8mm
10. plasterboard
11. gutter in transfer area(side average 80mm)
12. gutter(side 100mm)

详图1 detail 1

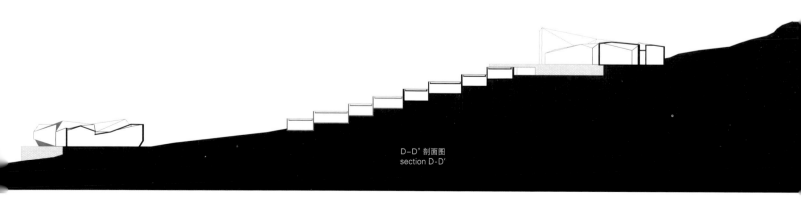

D-D' 剖面图
section D-D'

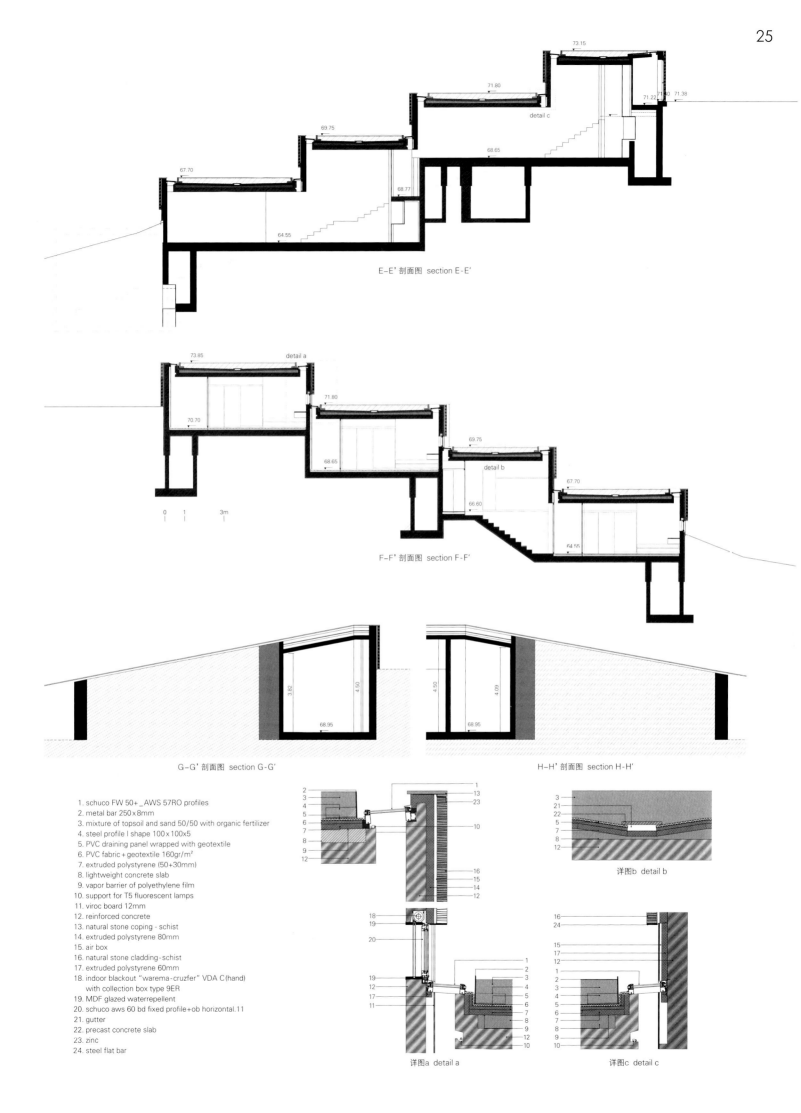

historical precedent to put the words in the mouth of the president". As regards the needs and wants of designing (as if architecture were not also a conscious act of will and innovation), they in turn also played out within "pre-existing" requisites(such as ensuring "Mobility and Accessibility for All", and the ever-present essential values of "Sustainable Development"), and those are materialized during the design process. Among those last the architects can emphasize the problem of taking on a large program (8,000m²/84 rooms/approx. 130 users), with the prospect of future expansion (up to 11,500m²/170 rooms/approx. 225 users) in a possible subsequent expansion phase of the housing area, without a significant impact on size and the landscape.

In the resulting interaction of the complex, the decision to structure the program in three fundamental components (Social Area, Room Area and Training Zone) merges with the (re-) interpretation of two elements of secular construction of the Douro landscape: the ubiquitous terracing, a recurring form of "inhabiting" this markedly sloping valley (here "inhabit" is read as "extracting bread from the earth"), and the large white bulks of the buildings set in the landscape among those stepped terraces. In particular the large wineries, are volumetrically and formally complex (often resulting from building over a long period of time, due to successive changes in the requirements of working on the land).

Between them (terraces and buildings – often between them and the river as well) the architects find abrupt, tense connections tearing through terraces, steep ramps, and stairs between walls, in order to meet the needs of the program.

The above is also an expression of the typical understanding of the history of architecture at the Faculty of Architecture of the University of Porto, not as an end in itself, but as one more element brought to the drawing board/computer, in coordination with other design problems.

As an engaging and exciting architectural challenge, the center was also a challenge in investigation of the forms and processes of the integration of the specificity of "new" themes, such as "accessibility" and "sustainability", in what the architects seek, the architects define, without adding adjectives that only lessen it. Nor "environmental", or "green", "accessible", or "sustainable". Architecture, the true architecture, for being it, is all that. And much, much more.

Álvaro Fernandes Andrade

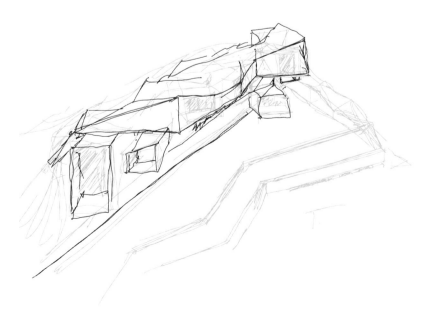

休闲小筑

古代社会之间的联系,主要依据的是简单的社会和经济结构间的彼此关联,这一点在当时的建筑中便展现得淋漓尽致。个体生命的日常生活主要包括劳动和睡觉,以致于享用休闲建筑的这一优待权只限于统治阶级。如果当时在工业革命的首要阶段中,于生产效率方面能进一步增加工人的独立性,再加上社会学说渐次所取得的成就,对普通大众来说,得到新的自由空间不再是难事,日益增加的休闲时光是与交通和信息手段(以进行更有效的信息传播和交流)的蔓延与经济性相联系的,且使更多的人们从遥远的城市中来到令人舒适的地方,享受到假期。一系列相互连接的功能性地形便是人类社会进化的产物,但是大部分在我们适应了日益增长的休闲时间(包括私人的和集体的)的需要时,还都处于人们的想象中。

Ancient societies were articulated according to simple social and economical structures that were mirrored in their architecture. The individual life course of an ordinary person consisted mainly of labor and sleep, so that architecture for leisure was a restricted privilege of the dominant classes. If in its first stage the industrial revolution further increased the dependence of workers on the rhythms of production, progressively, and thanks to the achievements of new social doctrines, new spaces of freedom were opened to common people. Associated with the spread of cheap means of transportation and of techniques for better information and communication, increased free time made it possible for ever larger populations to enjoy holiday periods in pleasant locations far from cities. A well-articulated range of functional typologies has been one product of this evolution of human society, but much remains to be imagined as we adapt to the growing demands for leisure time, both individually and collectively.

休闲住宅

在一开始的时候,人们有了比寻找简单的住所更为复杂的居住需求。精神上的这一居住需求对仅仅是一处栖身之所的住宅而言是截然相反的,山顶洞人如何通过手来标记墙壁或者通过在居住地标记狩猎圈来存储记忆便很好地阐述了这一道理。随着对住宅的需求相继而来,施工技术逐步创造出的纪念建筑为表达对世俗的非凡能力和对上帝的颂扬提供了强有力的可能性。

在很长的一段时间里,在主要的建筑领域中,住宅都是权力的象征以及被用作宗教实践的地方。休闲建筑则专门留给帝王和贵族使用,或者充当大型的公共设施,例如罗马浴场。之后,随着基督教的建立,这个存于典型罗马帝国社会中,穷人有权进去的集体休息场所也随之被瓦解了。

几个世纪以来,白天工作,晚上休息一直是个体生命的生活节奏。一周有六天,周天则被划分出去,用来履行宗教任务。对于极少数的贵族而言,他们拥有奢侈的闲暇时间,而对普通人来说则没有休闲的概念。仅有的宗教节日也都要从事艰苦的活动和进行集会庆祝,而不是让身体和心灵得到补给。

Free-time Dwellings

In the very beginning, mankind felt the need to dwell, a behavior much more complex than simple shelter-seeking. The mental attitude of dwelling, as opposed to mere sheltering, was well represented in how cave inhabitants marked its walls with imprints of their hands, or stored the memory of a hunt by inscribing its scenes in a living space. Building was a step that followed dwelling, whereupon the great possibilities of expression that construction techniques offered progressively led to the creation of monuments in celebration of secular powers and the gods.
Dwelling, displaying power and enabling the practice of religion was for a long time thus primary fields for architecture. Leisure architecture was reserved for emperors and aristocrats or for large public facilities such as the Roman baths. Incidentally, with the establishment of Christian society, this collective area, typical of Roman imperial society, accessible even to the poor, was eliminated. For centuries the course of an individual life was a succession of daytime work and nighttime sleep, six days a week, with Sunday set aside for the practice of religious duties. Idle time was a luxury reserved for a numerically tiny aristocracy, while the very concept of leisure did not exist for common people. The few religious holidays were devoted to strenuous activities or collective celebration and not to the individual regeneration of body and spirit.
Relatively simple societies produced a relatively uncomplicated

Timmelsjoch体验馆_The Timmelsjoch Experience/Werner Tscholl Architekt
山脊咖啡馆_Knoll Ridge Cafe/Harris Butt Architecture
帕德尔内的住宅_House in Paderne/Carlos Quintáns Eiras
福古岛的旅馆_Fogo Island Inn/Saunders Architecture
丽阿海滩度假住宅_Lia Beach Vacation Residence/Mold Architects
安布兰海滩俱乐部_Amburan Beach Club/Erginoğlu & Çalışlar Architects
维拉多姆斯儿童夏令营_Viladoms Children's Summer Camps/OAB
圣乔治侦察空间_Saint George Scout Space/Mutar Arquitectos
朝圣旅馆_Pilgrim Hostel/Sergio Rojo
休闲住宅_Free-time Dwellings/Aldo Vanini

相对简单的社会所造就的建筑类型相对的就没有那么复杂。随着建筑首次确定场地和建筑工人（受利益驱使涌入城市，通过满足人们日益增长的产品需求来赚取财富时），工业时代便带来了全新的社会和经济复杂性。在当今社会，此类建筑仍有存在的必要性。假定一个世纪以前，如果工业生产所积累的财富可以使更多人口，而且是超过那些仅仅需要生存的人口，有能力来安排时间进行娱乐和休闲活动，那么这些建筑便是必要的；与此同时，私人和公共交通设施的增多也直接便于人们在闲暇的时间里远离城市环境，纵情于户外运动，享受自然。

这一切使得旅游成为不仅限于贵族和上层阶级的活动。旅游事业的蓬勃发展催生了超于以往品牌宾馆和豪华假日别墅等大范围的度假胜地。然而，巨大的承载力并没有理所当然地暗示低质量的建筑；人们增加了大量智能且非常精致的建筑来供应人类需求，以用于满足人们足以负担的闲暇时间：智能的简易性、优越环境的融合、自然环保材料的创新性利用以及极简的设计，都被认为是更为先进的也是令人满意的高雅形式。

此外，这项新的事业本身不仅仅承载着来自经济上的压力，也承载了来自日益增长的人类需求同自然环境间可持续发展的敏感关系以及减少基于奢侈和无节制的生活方式所带来吸引力的压力。建筑满足了人们的需求，打造了更适于休闲的真实空间：帮助人们远离城市里的日常琐碎生活，重获与自然失联的慢生活节奏。考虑到当前休闲时间的日趋不足，这里所列举的所有例子，都能够通过合理安排自身和外部环境的简明关系，集中运用非人工材料，并且提供丰富的全景因素，来直接促进人与自然间的联系。

营造以同自然密切联系为导向的生活方式，树立团结一致和社会合作的意识，童军运动教育了年轻的一代儿女要高效地运用自由时间。座落于智利圣地亚哥的乡村，由Mutar建筑事务所设计的为圣乔治侦察空间所建造的项目运用现代语言解释了当代人们的侦查活动的必要性。两座平行的木覆层建筑向外延伸，尾端呈开放状态，用于从事主要的日常活动，并且在中间地带建造了一个带有屋顶的广场。除了地下室之外，没有再称得上是"屋子"的内部空间，学校和工作室被构思成通向外面的统一结构，同木质嵌板上的斜切设计相映衬，但木质嵌板却不是"窗户"。建筑采用同样简单却结实的材料，便于完成夏令营侦查活动。铁路枕木

range of architectural typologies. The industrial era brought a completely new social and economical complexity, with architecture for the first time addressing the worksites and residences of workers who had been drawn to the cities by the wealth that came from satisfying growing production needs. Such architecture remains necessary today, given that for more than a century the wealth accumulated by industrial production has allowed larger sectors of the population resources beyond those required for mere survival, along with the capacity to afford organized time for leisure and recreation. Meanwhile, the diffusion of individual and public transportation has directed free-time enjoyment away from the urban environment and toward open-air activities.

Travel has become an option not restricted to the aristocracy or upper classes, while a widespread tourist movement has generated a wide range of holiday resorts beyond the grand hotels and luxury holiday villas of the past. However, greater affordability does not necessarily imply lower architectural quality: Increasingly diffuse smart, highly sophisticated architectural provisions for more affordable free time solutions have emerged: Clever simplicity, superior environmental integration, innovative use of natural materials, and minimalist design are now considered more advanced and desirable forms of elegant sophistication.

What's more, this new course stems not merely from economic pressures, but from a growing sensitivity to the need for a more sustainable relationship with the environment and a reduced attractiveness of lifestyles based on luxury and excess. Architecture can serve these needs by creating spaces more suited to the actual essence of leisure: escape from the daily urban routine and rediscovery of lost contact with nature's slower tempo. In consideration of the increasing scarcity of leisure time, all the examples proposed here are able to enhance the immediacy of this contact by arranging precise relationships between interior and exterior, by extensively applying non-artificial materials, and by providing ample panoramic elements.

Pioneering a way of life oriented toward a close relationship with nature and promoting a great sense of solidarity and social cooperation, the Scout Movement has educated generations of boys and girls in the productive and responsible use of free time. The project of Mutar Arquitectos for a Saint George Scout Space in the Santiago de Chile countryside interprets the essential points of the scouting program in contemporary language. Two parallel wood-paneled volumes, open at the ends, concentrate common activities and generate an intermediate roofed square. With the exception of the cellars, no internal space can be considered a "room", and the school and workshop areas are conceived along a continuum towards the outdoors, accented with oblique cuts in wood panels that refuse to be treated as "windows". The same simple, solid materials are used for the buildings and for the fa-

福古岛的旅馆，纽芬兰岛，加拿大
Fogo Island Inn, Newfoundland, Canada

Timmelsjoch体验馆，Timmelsjoch，奥地利和意大利
The Timmelsjoch Experience, Timmelsjoch, Austria and Italy

和石笼石铺成的路面连接起了多处户外帐篷区和集体活动区。一个高于树梢的木质瞭望塔，除了起到保护作用，也体现了营地的非戏剧性安扎方式。

夏季，沉浸于大自然中，寓教于乐，成为一个最受人欢迎的度假方式。在西班牙野外的一个小镇，坐落于Castellbell i el Vilar，由OAB构想的维拉多姆斯儿童夏令营，可以同时允许儿童、成人和家庭度假。建筑物的周围都是白色的小屋，专门提供公共服务，其间的具有自然特征的教室让人们回忆起了典型的儿童对"房子"和"村庄"的想法。可持续性、耐用性、适应性和紧缩性是这个项目中的地基理念的关键词，统一且均匀覆盖体量的先进的高科技材料也易于维护和提高经济效率。

当代人们寻找便于消遣的空闲时间，沿着通往西班牙北部的圣地亚哥大教堂开始了朝圣之旅，继续追求一种怀旧的体验，追逐废弃的以醇厚古老的艺术和文学著称的李奇欧音乐学院。在那里，Sergio Rojo利用当代的优雅语言，通过其当代基本的住宿和服务来诠释传统的朝圣旅馆。结构的空间观念同朝圣者体验的精神追求和谐一致。朝圣者旅馆内部没有装饰，以白墙为主导，且放置了斯巴达式的简单家具，安装了可见的木质屋顶，这一切都符合了遵循着古老的圣地亚哥之路的德孔波斯特拉古城的神秘主义。在这里，由于虔诚的休憩活动的进行，因此人类生活的痕迹得以保留。

奢华并非必须暗示材料要多么的昂贵，装饰多么的繁琐。独特的地理位置、优雅合理的设计都可以被看作是更可取的奢华形式。

在利用独特的选址这一相同的概念下，由Shorefast基金会委托的纽芬兰岛福古岛的旅馆展现了更为复杂的项目，以促进地区文化生产力和地区旅游业的发展，来为当地经济注入活力。旅馆还提供了一个艺术长廊、私人电影院、健身房、图书馆和会议室，同客房设置在一起。Saunders建筑事务所将这座旅馆设计成X形布局的两个体量，四层楼，建于细长、成角的钢柱之上，连接起来，使得整栋建筑远离高低不平的地面，且感受不到太平洋海浪的激荡。坚固的几何体量是由木板以及一应俱全的窗户组成的，最终形成了多层餐厅的大型玻璃墙，面向大海开放。即使建在原始荒落的野外，它的存在也显不出有丝毫的侵入感。偶然途经的游人们惊叹它的美，酒店的幸运房客们也赞美着这美丽的风景。

现代都市人每天都要面对着专业技术的问题，他们可能在临时的、

cilities that complete the scout camp. Railway sleepers and gabion stones form the paths connect the various outdoor areas for tents and collective activities. A wooden watchtower stands above the treetops and, in addition to its utility for safety, indicates the camp's presence in a non-dramatic way.

A summer education immersed in nature is becoming a favorite way to spend holiday time. The OAB conceived Viladoms Children's Summer Camps in Castellbell i el Vilar, Spain, as a small town immersed in the wild, able to accommodate children, adults and families. The white cabins arranged around the buildings of the common services and of the natural classrooms recall archetypical, childhood ideas of "house" and "village". Sustainability, durability, adaptability, and austerity are keywords of the Foundation philosophy underlying this project, and the advanced hi-tech material uniformly cladding the volumes enhances both ease of maintenance and economic efficiency.

In the search for well-spent free time, contemporary people continue to pursue such old-fashioned experiences as the devotional pilgrimage to Santiago de Compostela in Northern Spain, a quest which starts from an abandoned Liceo, an ancient school for arts and literature. There, Sergio Rojo interprets the traditional Pilgrim Hostel, with its basic accommodations and services, in contemporary, elegant language. The structure's spatial perceptions are in perfect harmony with the spiritual purpose of the pilgrimage experience: The absence of decoration, the dominant whiteness of the walls, the Spartan simplicity of the furnishings, and the visible wooden structures of the roof suit the mysticism of those who follow the ancient Camino de Santiago de Compostela. Here the time of human affairs is left behind, for a very particular kind of devotional otium.

Luxury doesn't imply necessarily expensive materials or overdecoration. An exclusive location and an elegant, rational design can be considered a more desirable form of luxury.

Under the same concept of exploiting an extraordinary site, but articulated in a more complex program, Newfoundland's Fogo Island Inn was commissioned by the Shorefast Foundation to enhance cultural production and geo-tourism and to energize the local economy. Along with guest rooms, the Inn offers an art gallery, a movie theater, a gym, a library and meeting rooms. Saunders Architecture designed the inn as two X-crossed, four-storey volumes standing on a forest of slender, angulated steel trunks that insulate the building from the rough terrain and the force of the Atlantic surf. The strong, geometrical volumes are finely textured by wooden panels and a rich repertoire of windows, culminating in the huge ocean-facing glass wall of the multilevel restaurant. Even if the building rests on a virgin, wild place, its presence

维拉多姆斯登儿童夏令营，Castellbell i el Vilar，西班牙
Viladoms Children's Summer Camp, Castellbell i el Vilar, Spain

帕德尔内的住宅，卢戈，西班牙
House in Paderne, Lugo, Spain

向天堂一样的地方寻找答案，这意味着建筑可以把人带回到过去简单的乡村生活当中。由Carlos Quintáns Eiras设计的、坐落于西班牙的帕德尔内的住宅作为对老旧的乡村谷仓的重塑，模仿了原始谷仓的外形，并且利用其绿色斜坡上的地理位置，赋予室内一处现代感十足且精致的环境，同时展现了乡村的优美风景。建筑的一部分建在谷仓的古老的石头墙上，覆层为胶合板，嵌在砌体石墙上，以一个高雅的形式同乡村环境融为一体。类似的休闲建筑的构思大体相同，作为一种优先的、外部的角度观察视点，不一定能融入到环境中，这些在下面的例子中得到了很好的体现。

Werner Tscholl建筑师事务所设计的Timmelsjoch体验馆便采用了相同的方法，从壮观的天文台可以俯瞰意大利的布伦纳山口。这个天文台结构是建筑雕塑系统的一部分，它延伸至山谷，为途经的游客们提供沉思体验的机会，让人联想到信奉浪漫主义的跋涉者à la Caspar David Friedrich。

面对尚未被污染的自然，令人舒适的沉思看起来是衡量Harris Butt设计的山脊咖啡馆的一个要点，以满足享受休闲时光的渴望。

建筑师采用了不同以往的、更为复杂的方法来处理与场地之间的关系，这一特点在Mold建筑事务所设计的丽阿海滩度假住宅中有所体现。该项目寻求能够最大化地整合爱琴海迷人的风景同建筑之间的关系。在非透明性这一点上，建筑元素和设施有所改造，使其成为风景的一部分，这在第一个石头公路案例以及第二个非凡的屋顶游泳池案例中都有所展示。

通过对比，Erginoğlu & Çalışlar建筑师事务所重塑了印度尼西亚安布兰海滩俱乐部的传统休闲设施，将注重独立的内向型风景的发展，没有继续以海滩和游泳池为中心，而是分散了限定的建筑元素和城市里的便利设施，从而将整个风景名胜区带入了新的高度。

这里所讨论的案例的类型跨度极大，反映了大量的休闲建筑概念。如果有一条主线贯穿其中，这主线可能是景观在视觉方面所产生的娱乐性的调查，以及建筑作为观赏点，在一定程度上脱离景色，反映了人们对于旁观者（而非主角）所处的风景的不同态度的方式。

doesn't appear intrusive, like that of an occasional, passing visitor who enjoys, as lucky guests in the inn, the landscape's beauty. Breasting daily confrontations with technology and professional struggles, the contemporary urbanite may seek answers in the temporary havens of places and buildings meant to bring one back to the simplicity of the past and the countryside. Designed by Carlos Quintáns Eiras as a remodel of an old country barn, the House in Paderne, Spain, mimics the external shape of the original destination, offering inside a modern, sophisticated environment and a beautiful view of the countryside that takes advantage of its position on a green slope. Built partly on the barn's ancient stone walls, the house is finished with laminated wood mounted on masonry walls in an elegant integration with the rural surroundings. A similar tendency to conceive of leisure architecture as a privileged but external point of observation, not necessarily integrated into the context, is well represented in the following examples.
An identical approach is followed by Werner Tscholl Architekt in the Timmelsjoch Experience, a spectacular observatory overlooking Italy's Brenner Pass. The structure is part of a system of architectural sculptures, and it protrudes into the valley offering passing tourists a sublime experience of contemplation, reminiscent of the romantic wayfarers à la Caspar David Friedrich.
A detached, comfortable contemplation of uncontaminated nature seems to be a key point for the Knoll Ridge Cafe by Harris Butt, in satisfying the desire for an occasional fruition of leisure.
A different and more complex approach to the relationship with the site characterizes the Lia Beach Vacation Residence by Mold Architects which seeks maximum integration with the stunning landscape of the Aegean Sea. It does so to the point of a kind of invisibility, by transforming architectural elements and facilities into parts of the landscape itself, as shown in the stone-delimited paths in the first case and the extraordinary roof-pool in the second.
By contrast, Erginoğlu & Çalışlar Architects, remodeling the old-fashioned leisure facility of Indonesia's Amburan Beach Club, focused on the development of a self-contained introverted landscape, and no longer centered on the beach and the pool, but dispersed its qualifying architectural elements and amenities all around the village, thus bringing the entire resort to a new fruition.
From the examples discussed here, one can conclude that architecture for leisure spans quite a wide range of typologies, reflecting an equally wide range of concepts of leisure. If a common thread runs among them, it may be the research for enjoyment, essentially visual, of the landscape, and how the architecture thus presents itself as an observation point, remaining in some way apart from the scene, revealing the diffuse attitude toward the scenery of spectator rather than protagonist. Aldo Vanini

Timmelsjoch体验馆

Werner Tscholl Architekt

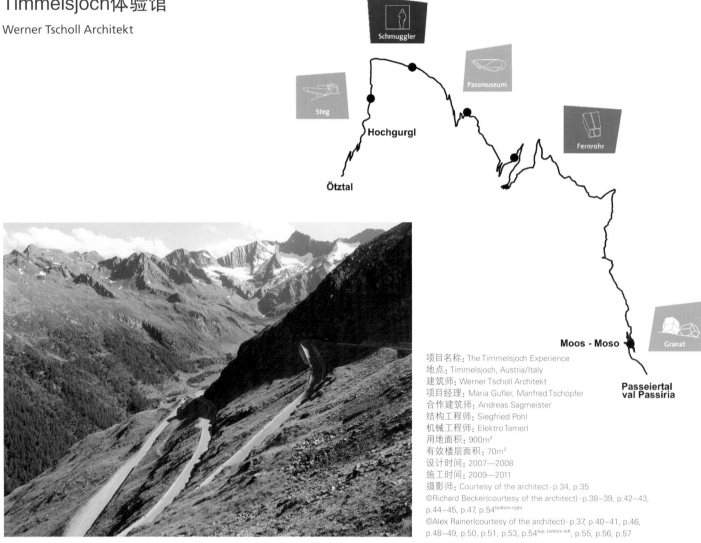

项目名称：The Timmelsjoch Experience
地点：Timmelsjoch, Austria/Italy
建筑师：Werner Tscholl Architekt
项目经理：Maria Gufler, Manfred Tschopfer
合作建筑师：Andreas Sagmeister
结构工程师：Siegfried Pohl
机械工程师：Elektro Tamerl
用地面积：900m²
有效楼层面积：70m²
设计时间：2007—2008
施工时间：2009—2011
摄影师：Courtesy of the architect - p.34, p.35
©Richard Becker(courtesy of the architect) - p.38~39, p.42~43, p.44~45, p.47, p.54 bottom-right
©Alex Rainer(courtesy of the architect) - p.37, p.40~41, p.46, p.48~49, p.50, p.51, p.53, p.54 top, bottom-left, p.55, p.56, p.57

  Timmelsjoch山坐落于雷申山口和布伦纳山口之间，是阿尔卑斯山主峰上最深的、没有冰川痕迹的山口。

  曾经只有一条羊肠小道蜿蜒在Passeiertal山谷与厄兹塔尔山谷之间。这条古道不仅利于贸易的进行，同时也在社会、文化、政治和宗教方面举足轻重。那个时候，大车道屈指可数，因此旅者、小贩和赶驮畜的人们宁可选择最短的路径也不会选择最容易行走的路线。

  自1968年以来，一条拥有数个回头弯的古道将这些山谷以及奥地利和意大利两个国家连接起来。这个庞大的技术工程在54年间通过Timmelsjoch体验馆进行了扩展。Timmelsjoch体验馆是Timmelsjoch Hochalpenstraße AG (Passeiertal山谷以及南蒂罗尔省和提洛尔省内的一个Moos公社) 资助的一个文化项目。

自2010年开始,司机们驾车行驶在Timmelsjoch公路上,穿梭往返于Obergurgl和Moos时,可以观赏到五座截然不同的建筑雕塑,它们的形式与色彩源于周围的景观。这些雕塑由染色的混凝土构成,其材料连同木质框架为室外墙体赋予了特色。内部则是根据展览的主题,采用了彩色的印刷玻璃,来呈现本地域的自然、历史、文化、社会和经济信息。建筑造形的设计因素同地形以及周围的景色和谐相融。

Timmelsjoch体验馆的体验始于意大利境内的石榴石,它们伫立在岩石坡上,利于全方位俯瞰从Moos村庄绵延到Passeiertal山谷的绝佳风光。在Passeiertal山谷发现的两块石榴石是地理岩石形成的典型之作,用作展览室和观景台。

而处在意大利一侧的山口,人们的视野集中在Granatkogel山(3304m高)和Hohe First山(3403m高),它们醒目地伫立在这片永久的冰川之上。Scheibkopf山下广袤无垠,为欣赏Texelgruppe自然保护区的全景提供了180度的豪华视角。

在道路的最高点上,北蒂罗尔省一侧的山口博物馆的混凝土结构像一块巨石一样伸向南蒂罗尔省一侧,突出了Timmelsjoch体验博物馆里交相辉映的特征。博物馆里的"冰穴"则是向建造阿尔卑斯高山公路的先驱们以及他们所取得的伟大成就致以敬意。

在Timmelsjoch公路横穿古道的地方,建筑师建造了可供人参观的立方体建筑。古道一直从Zwieselstein通往Passeiertal的Moos区,展现了奇幻与冒险的世界。再现了走私贩们穿过Timmelsjoch古老路线,并且最终以奥地利古道(全然是新奇而又激动人心的风景)作为起点的场景,古道还横跨至厄兹塔尔山谷的尽头。在这条路上,令人惊叹的3000m高的层峦叠嶂也在全面描述着周围区域及其与众不同的特色。

The Timmelsjoch Experience

The Timmelsjoch is the deepest, non-glaciated indentation in the main Alpine ridge between the Reschen Pass and the Brenner Pass.

Once only a mule track linked the Passeiertal Valley and the Ötztal Valley. The ancient path not only did facilitate trade, but also was of great social, cultural, political and religious significance. At that time, cart tracks were few and far between, therefore travellers, pedlars and people leading pack animals didn't choose the easiest route, but rather the shortest one.

Since 1968 both these valleys and the countries of Austria and Italy have been connected by a road consisting of numerous hairpin bends. This gigantic technical project, meanwhile 54 years old, has been extended via the Timmelsjoch Experience, a cultural project supported by the Timmelsjoch Hochalpenstraße AG, the commune of Moos in the Passeiertal and the provinces of the South Tyrol and the Tirol.

Since 2010 between Obergurgl and Moos drivers on the Timmelsjoch Road are able to view five different architectural sculptures, their form and colour being oriented to the surrounding landscape. They consist mainly of dyed concrete, and its materiality, together with the markings of the wooden formwork, characterizes the outside walls. Inside colored printed glass is used according to the subject of the exhibition, providing information on the nature, history, culture, society and economy of this region. Designations are given to the buildings since the formative elements

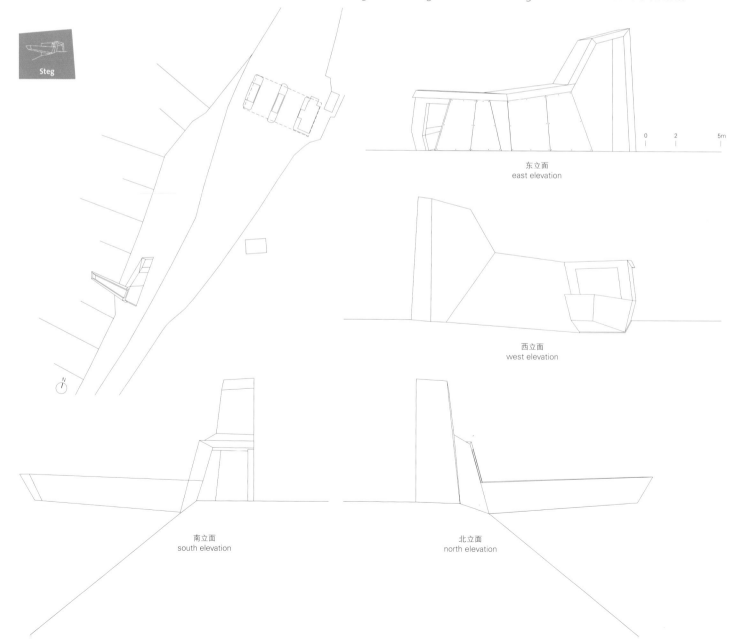

东立面
east elevation

西立面
west elevation

南立面
south elevation

北立面
north elevation

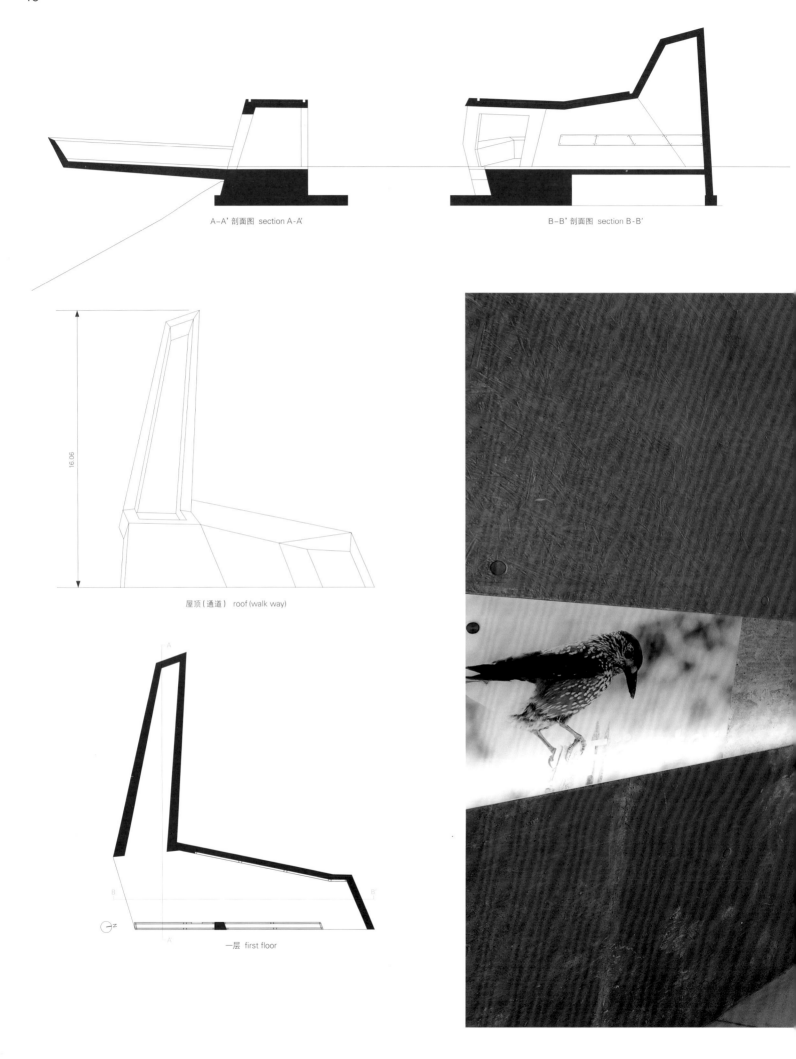

are related to the landscape and to the topography of the place. The experience starts in Italy with the garnets, standing on the rocky hillside, providing a stunning view down to the village of Moos and across the far end of the Passeiertal Valley. Two garnets, modelled on the geological rock formations found in the Passeiertal Valley, serve as an exhibition room and a viewing platform. Also on the Italian side of the Pass the telescope focuses the viewer's attention on the Granatkogel (3,304m) and the Hohe First (3,403m) which stand out prominently from the eternal glacial ice. The spacious area beneath the Scheibkopf Mountain offers a superb 180° panorama view of the Texelgruppe Nature Reserve.
On the highest point of the road the concrete structure of the Pass Museum on the North Tyrolean side juts out like an erratic boulder into the South Tyrolean side, underlining the cross-border nature of the Timmelsjoch Experience. The "Ice Cave" inside the museum pays tribute to the pioneers of the High Alpine Road and their remarkable accomplishment.

The walk-in cube created at the spot where the Timmelsjoch Road crosses the ancient trail that leads from Zwieselstein to Moos in Passeiertal, shows the world of adventure and danger, the world of the smugglers and their ancient route over the Timmelsjoch, and finally at the beginning of the road in Austria the walkway offers a totally new and utterly breathtaking perspective, across to the far end of the Ötztal Valley. Its awe-inspiring proliferation of 3,000-metre peaks, provides in-depth information about the surrounding area and its distinguishing features.

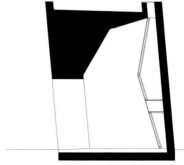
C-C' 剖面图 section C-C'

D-D' 剖面图 section D-D'

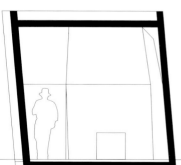
E-E' 剖面图 section E-E'

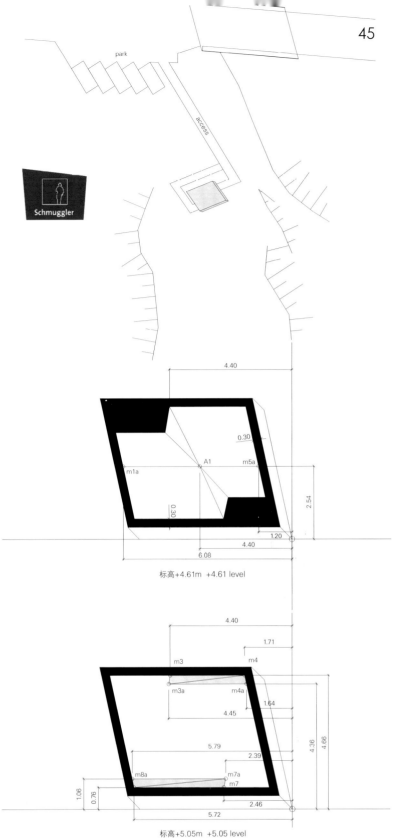

标高+4.61m +4.61 level

标高+5.05m +5.05 level

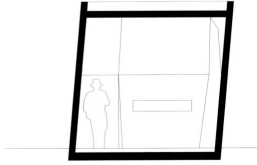

F-F' 剖面图 section F-F'

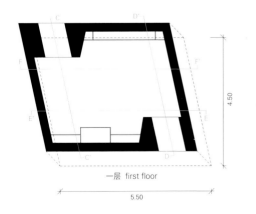

一层 first floor

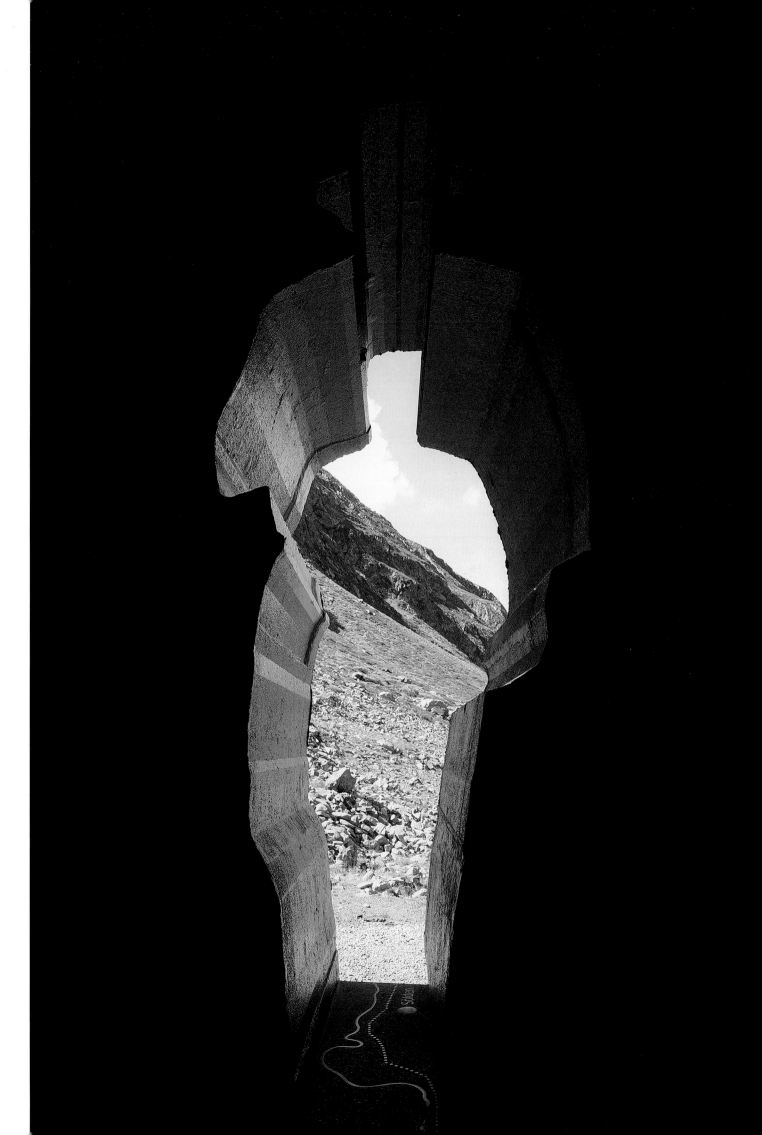

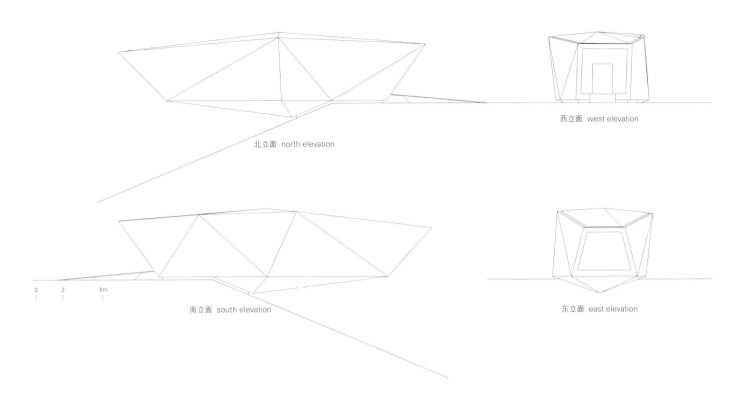

北立面 north elevation 西立面 west elevation

0 2 5m

南立面 south elevation 东立面 east elevation

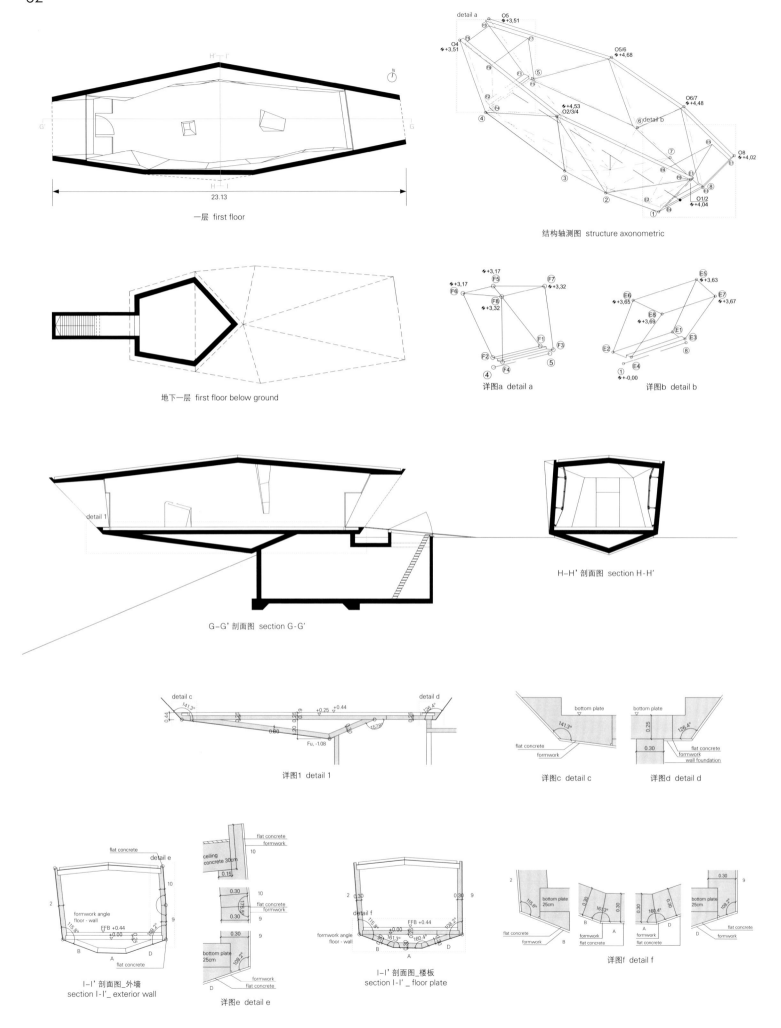

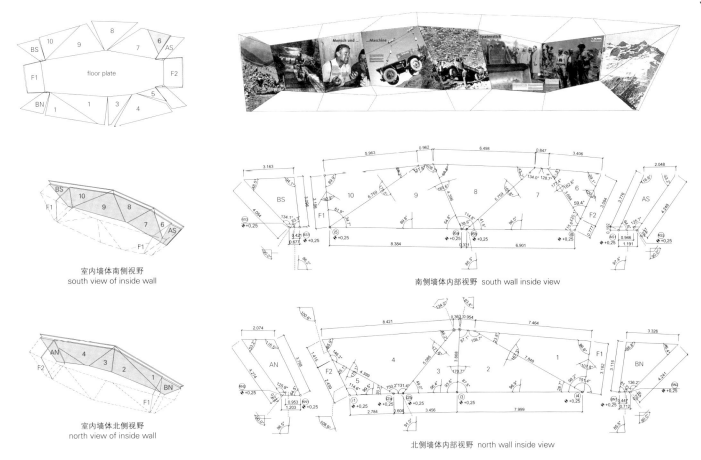

室内墙体南侧视野
south view of inside wall

南侧墙体内部视野 south wall inside view

室内墙体北侧视野
north view of inside wall

北侧墙体内部视野 north wall inside view

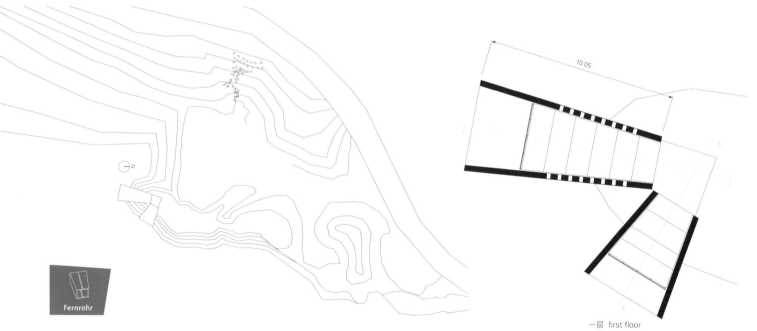

Fernrohr

一层 first floor

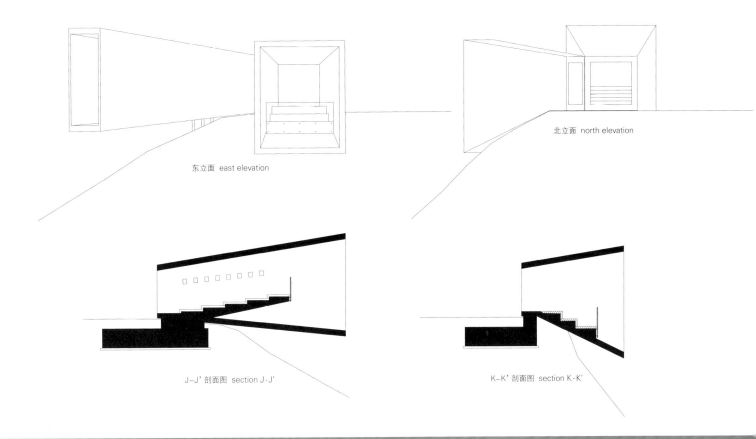

东立面 east elevation

北立面 north elevation

J-J' 剖面图 section J-J'

K-K' 剖面图 section K-K'

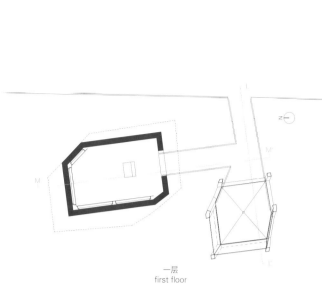

一层
first floor

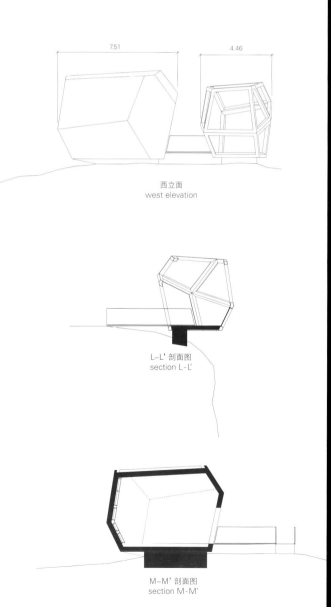

西立面
west elevation

L-L' 剖面图
section L-L'

M-M' 剖面图
section M-M'

山脊咖啡馆
Harris Butt Architecture

山脊咖啡馆座落于汤加里罗国家公园里鲁阿佩胡火山上的Whakapapa滑雪场内，这个商业滑雪场位于山体的一侧，同时也位于新西兰最大的活火山之上。

2009年2月，原址上的山脊小屋遭遇一场大火，面目全非，山脊咖啡馆便在原址内取而代之。2009年的冬天，作为暂时的替代山脊小屋的预制设施，这一宏伟的建筑项目开始实施。移除了原来山舍的残骸，220㎡的临时咖啡馆在存留的楼板上笔直而立。这个方法论经受住了考验，后来被应用于该咖啡馆的建造当中。

突变的天气是新西兰山脉的典型气候特征，鲁阿佩胡火山也不例外。如此极端的环境为建筑设计增添了独特的挑战。通往场地的有限的道路以及严苛的要求意味着要将材料运往施工地段需要大量的计划和物流。预制的混凝土楼板要在雪融化之前快速建造，及时配送，以便在下一年的夏季建筑工程开始前，就已经被运送到700m高的雪山之上。

建筑设计中主要考虑的是位置偏僻这一问题。整座建筑，从基座梁/楼板到屋顶以及窗户部分都被分解成预制模块系统，便于直升机在场地内搬运、放置和树立。咖啡馆的大部分墙体和屋顶采用的是由胶合板和单层板组成的保温夹层板。像大部分的建筑设计中的建筑零部件一样，这些部件的使用都谨慎地考虑到了直升机的最大载重为800kg的限制。

建筑师设计了一个100%隔热的玻璃幕墙，应用在可能是最具挑战性的环境中。玻璃和框架系统可以承受时速高达200km的风力以及零度以下的气温。虽未测量场地，但经估算，建造524㎡的玻璃墙预定的玻璃需求量为25t。所有的玻璃单元都配备了三条平衡管，来促进现场的氩气体填充，平衡管也作为预警机制，监控飞机飞行巾快速地高地加速。

在夏季的时候，极目远眺，人们可以看到建筑的东面坐落于火山石岩层之上，即Te Heuheu山谷下行的边缘。北面回望整座山，而西侧设有升降椅和滑雪区。建筑的形式强烈地反映了当地山脉的地理特点，在山顶处，形如"鸥翼"的屋顶像极了装载着山峰的"摇篮"。从实践的层面上来讲，这个设计起到了处理积雪的作用，整座建筑的目的是用于掩盖3m高的积雪。

建筑没有采用传统的形式，而是广泛使用木材，应用在内部和外部，从而营造传统山中小屋的温馨之感。外部的玻璃（尤其是东面墙体）是建筑的另一大特色，并且完全地置身于顶峰山脊的壮丽风景之中。

Knoll Ridge Cafe

Knoll Ridge Cafe is located at Whakapapa Ski Field on Mt. Ruapehu, Tongariro National Park. Situated on the side of a mountain the commercial ski field is also sited on what is New Zealand's largest active volcano.

The cafe replaces the original Knoll Ridge Chalet which was destroyed by a fire in February 2009. As a result an ambitious design was initiated to replace the chalet with a temporary prefabricated facility for the 2009 winter season. Once the debrises of the original chalet were removed, a 220sqm temporary cafe was erected on the remaining floor slab. This tested the methodology which was later adopted for the construction of the cafe.

项目名称：Knoll Ridge Cafe
地点：Whakapapa Ski Field, Mt. Ruapehu, Tongariro National Park, New Zealand
建筑师：Grant Harris
项目团队：Grant Harris, Ian Butt, Kerry Reyburn, Ben Brown
总建筑面积：1,516m²
竣工时间：2011
摄影师：
©Simon Devitt(courtesy of the architect)(except as noted)

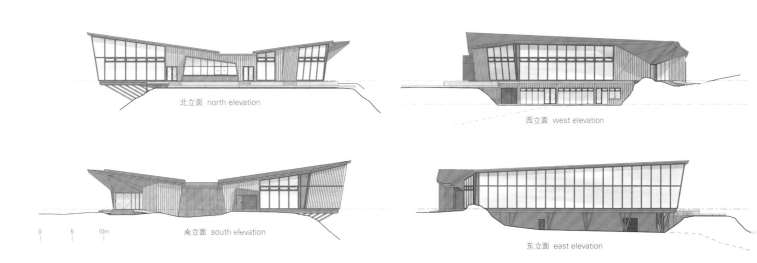

北立面 north elevation   西立面 west elevation

0 5 10m

南立面 south elevation   东立面 east elevation

1 大堂 2 会议室 3 服务区 4 休息室 5 清洁室
6 储藏室 7 员工更衣室 8 员工室 9 货柜 10 通道
1. lobby 2. meeting 3. service 4. restroom 5. cleaner 6. store
7. staff locker 8. staff room 9. container 10. passage

地下一层 first floor below ground

1 入口大堂 2 更衣室/大堂 3 咖啡馆 4 备餐间 5 咖啡馆/酒吧 6 厨房 7 干燥室
8 冷冻室 9 冷却装置 10 办公室 11 休息室 12 清洁室 13 洗涤槽 14 接种室 15 平台
1. entry lobby 2. locker/lobby 3. cafe 4. servery 5. cafe/bar 6. kitchen 7. dry store 8. freezer
9. chiller 10. office 11. restroom 12. cleaner 13. dishwasher 14. transfer room 15. deck

一层 first floor

Rapidly changing weather is typical of the conditions encountered on New Zealand mountains, with Mt. Ruapehu no exception. Designing a building for such a severe environment provided its own set of unique challenges. Limited road access to site and stringent requirements meant extensive planning and logistics were required just to get materials to site. Prefabricated concrete floor panels had to be rapidly constructed and delivered before the snow melted, these were then hauled over snow 700m up to site before construction began in the following summer.

A major consideration in the design of the building was the issue of the remote location. The entire building, from foundation beams/floor panels to roof sections and windows was broken down into a modular panelised system, which allowed for delivery, placement and erection by helicopter on site. Insulated sandwich panels constructed of plywood and LVL form a large extent of the walls and roof of the cafe. These like most of the building's components had to be designed with a careful consideration of not exceeding the helicopter's 800kg max load limit.

A 100% thermally glass curtain wall was designed for what is possibly one of the most challenging environments to build in. The glass

and framing system had to withstand wind speeds of up to 200km/ph and temperatures well below freezing. Twenty-five tons of glass was used in the 415m² area of glass facade which was all predetermined and ordered from calculations without a site measure. All the glass units were fitted with 3 equalizing tubes to facilitate on-site argon gas filling, and equalizing tubes were also used as a precaution for rapid altitude acceleration during flight.

In the summer season the eastern face of the building can be seen set above the volcanic rock formations located on the edge of the drop-off to the Te Heuheu Valley. The north face looks back down the mountain whilst to the west is the chair lift and ski area. The form of the building reflects the strong geological features of the mountain. The "gull wing" roof was to appear to "cradle" the mountain's peak. On a practical level it is used to manage the snow. The building is designed to cover with up 3.0m of snow. Timber has been used extensively inside and outside to create the warm "feeling" of the "traditional" mountain chalet without adopting the traditional form. The exterior glass (particularly to the east wall) is the other feature of the building – allowing full exposure to the magnificent view to the Pinnacle Ridge.

帕德尔内的住宅
Carlos Quintáns Eiras

　　O Courel山脉是处别具一格的地方。它保留了一些由于开发以及一些自然景观的更替和标准化演变而消失的一些特色。它还保留了另一个时空，另一种韵味，另一种色彩，保留了稀有的植被和原汁原味的村庄，帕德尔内便是其中的村落之一。在这处人杰地灵的地方，一个古老的谷仓坐落于此，而一座现代化建筑位于其上。

　　谷仓增建的部分已经拆除，而古老又厚重的石头墙被很好地保留下来，作为新木质结构的基础。一层是卧室，顶层是观景台，类似一块高地，可瞻望地形景观。

　　石砌墙利用其厚重感对卧室提供了保护，同时木结构则保护了生活，展现了两种不同的建筑方式。这样的房子通透可见，没有意在隐藏任何事物。顶层使拐角尽可能的多，实现了上下调节，从而满足了优化空间的目的。

　　在楼上，大同小异的木质结构变换着形状，来搭配原始的墙体以及更多的木结构设计。而在楼梯下方，木材覆盖了卧室空间，赋予睡眠一种温暖感。楼上是一处大型开放场地，带有一个大型开口。

　　房间为螺旋状的布局，出口位于正中间，厨房位于最高点，卧室则位于最低点。顶楼的连续空间被分割成两个不同的层面，紧密地靠着出口，因此可以展现风景的所有魅力。分割这片区域意味着把这处空间改造成小房间的增建结构，所以建筑师所面临的挑战在于在没有物理界限的区域内建立有界限的功能区。这座住宅利用其尽可能小的规模，来与

每个角落相契合，并且凭借着其最高部分的空间来产生舒适感，也为日常使用平添了几分庄严感。

建筑结构由层压木板制成，嵌装在石墙上，楼层结构也是木质的。石板瓦屋顶让人们想起了村庄的古老建筑。屋顶采用了最大规格的石板瓦，其带有粗糙纹理的立面唤醒了几个世纪以来，在这个地方运用的传统的板栗树材质的木质立面。而在室内，木质MD胶合板填充了结构的缝隙。即使这座住宅采用了传统的材料，但是抽象的施工进程还是赋予了材料一种全新的外观。屋顶和立面之间的变化朴素而华丽。

House in Paderne

O Courel Mountain Range is a special place. It preserves some characteristics that have vanished because of development, transformations and standardisation of natural landscapes. It preserves another time, other flavours, other colours, exceptional vegetation and some untouched villages. Paderne is one of those villages. In this outstanding place, an old barn was located with a modern construction atop.

The new addition was demolished and ancient thick stone walls were preserved and used as a base for the new wooden structure. The ground floor houses bedrooms and the top floor is a view-

项目名称：House in Paderne
地点：Paderne, Folgoso do Caurel, Lugo, Spain
建筑师：Carlos Quintáns Eiras
项目团队：María Olmo Béjar, Borja López Cotelo
结构工程师：Francisco Carballo, Carolo Losada
承包商：Carlos López
甲方：Ángel Baltanás
建成面积：96.58m² / 使用面积：61.48m²
造价：EUR 110,000
设计时间：2008 / 施工时间：2009.6—2010.2
摄影师：©Ángel Baltanás (courtesy of the architect)

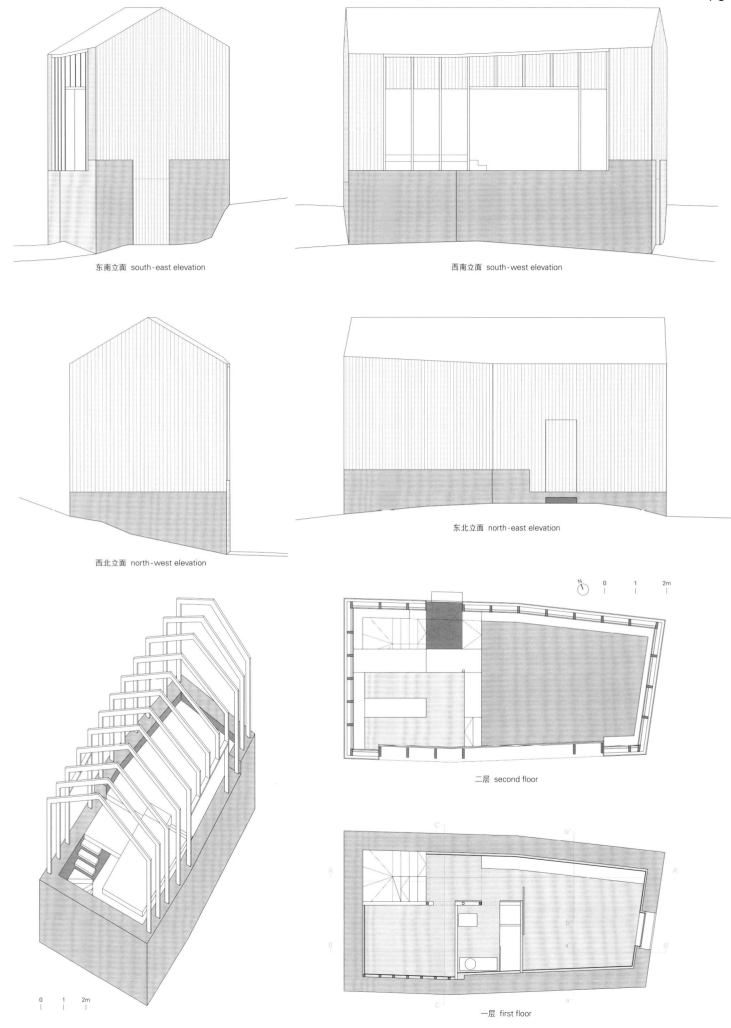

东南立面 south-east elevation

西南立面 south-west elevation

西北立面 north-west elevation

东北立面 north-east elevation

二层 second floor

一层 first floor

point, a plateau to look at the topographic spectacle.
Masonry walls shelter bedrooms in their thickness while wooden structure shelters life, showing both two ways of constructing. This house doesn't aim to hide anything, and everything is evident. The top floor makes the most of every single corner and goes up and down in order to optimize its spatial possibilities.
Upstairs, the wooden structure is repeated with little variations, being deformed to suit the original walls, and is accompanied by more wood. Downstairs, wood covers and gives the sleeping spaces warmth. Upstairs, there are a big open space and a big hole.
The house has a spiral organization and the access is located in the intermediate point. The highest point is the kitchen, and the lowest is bedrooms. The continuous space of the top floor is fragmented into two different levels that grant intimacy to the main access thus allowing to show the whole power of the landscape.

Compartmentalizing would have meant transforming the space into an addition of little rooms, so the challenge was to establish programmatic limits without physical limits. The house, with its minimum dimensions, suits every corner and relies on its high section of the spatial generosity that makes it comfortable and provides a certain grandeur in its daily use.

The structure is made of laminated wood mounted on masonry walls, and floor structures are wooden as well. The slate roof is a reminiscence of the ancient buildings of the village, and takes from its big size, while the rough texture of its facade evokes the traditional chestnut tree facades built over centuries in this place. Inwards, wooden MD plywood panels are placed into the gaps of the structure. Even if traditional materials have been used in the construction of the house, the abstraction process grants them a new way of looking, with the rotundity of plain changes between roof and facade.

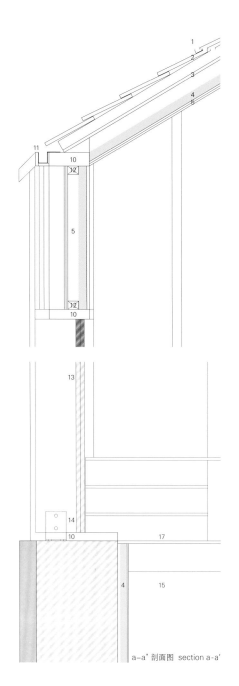

1. slate
   existing traditional slate from the former house
   variable thickness
   fixed with nails following traditional slate roofing
   from northern spain
2. timber battens, 60×72mm
3. plywood, 20mm
4. thermal insulation, rockwool panel, 60mm
5. double plywood layer, 20mm each
   nailed to main structure
   interior finishing, painted in white
6. beam reinforcement, timber beams, 950×700mm
7. timber beam, 100×210mm
8. wood coating, plywood, 20mm
9. waterproof membrane
10. timber frame, 300×950mm
11. zinc gutter fixed to timber frame
12. timber frame battens, 80×70mm
13. double glazing, 30mm each
14. steel plate fixation of timber structure
15. plywood layer, 20mm each
    nailed to main structure
    interior finishing, painted in white
16. existing stone walls
    variable thickness 600mm average
17. plywood, 20mm

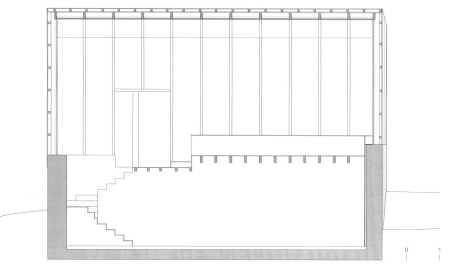

A-A' 剖面图 section A-A'

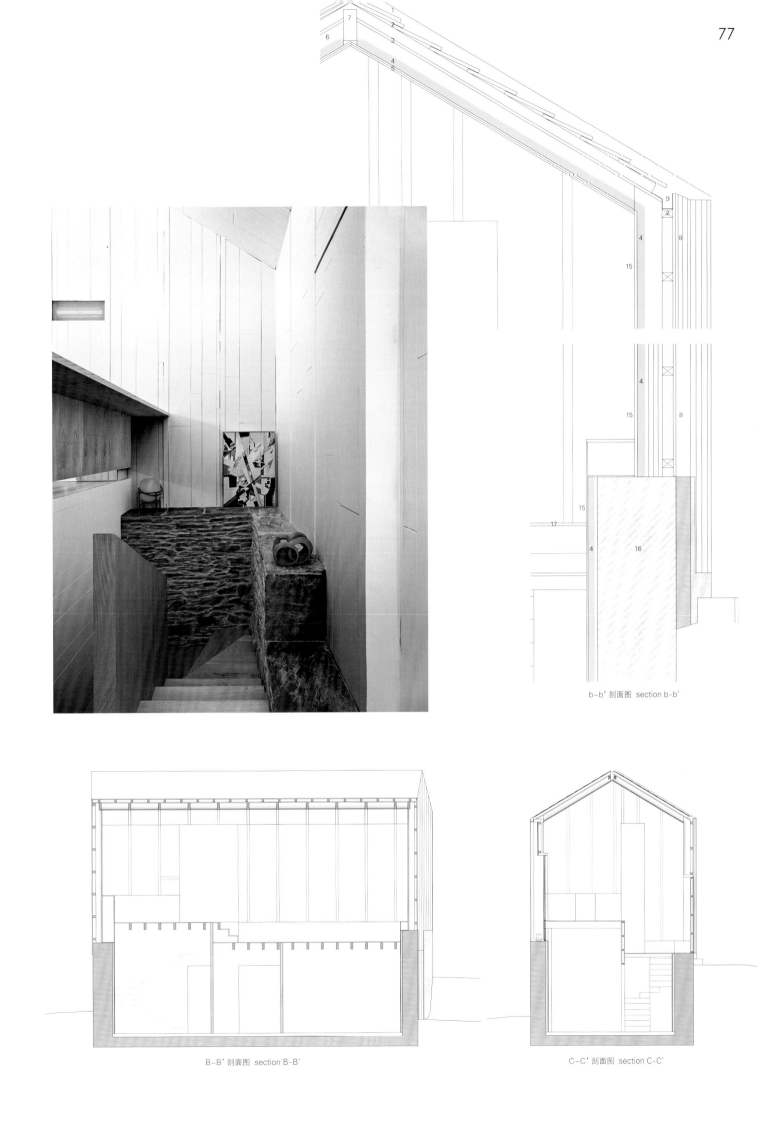

b-b' 剖面图 section b-b'

B-B' 剖面图 section B-B'

C-C' 剖面图 section C-C'

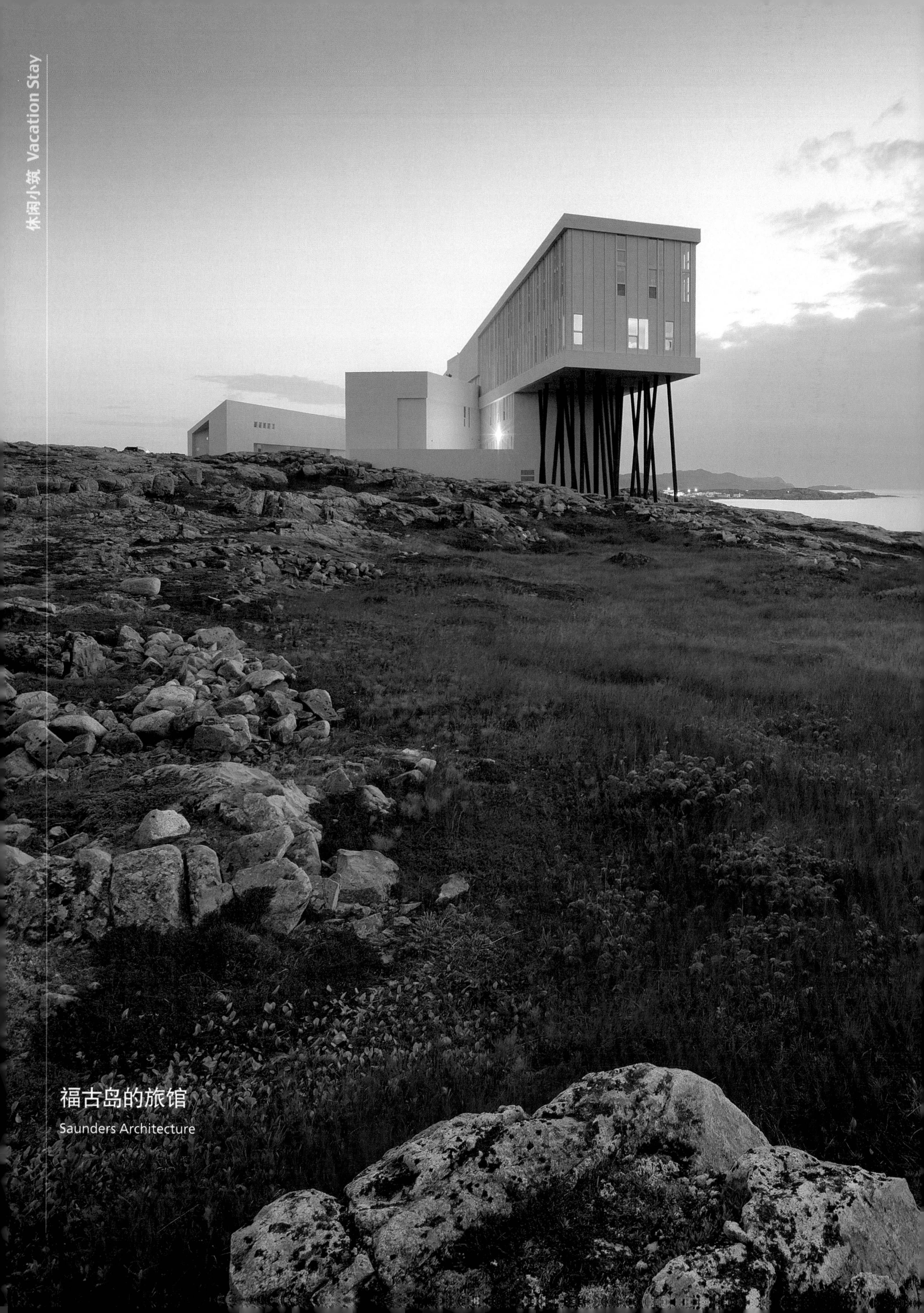

休闲小筑 Vacation Stay

福古岛的旅馆
Saunders Architecture

福古岛的旅馆是由Shorefast基金会构思的一个项目。Shorefast基金会是由Zita Cobb、Anthony Cobb和Alan Cobb领导的加拿大慈善机构。福古岛的旅馆作为一座研究性建筑，位于纽芬兰岛的东北部海岸，问世四个世纪以来，有助于将过往的历史再现于将来。

这座旅馆是加拿大最古老的定居地之一：致力于成为福古岛的文化和经济引擎，也回应了寻找传统知识和传统方式之间的联系的迫切需求。项目的目标是"利用古老的事物来发现新的方式"。几个世纪以来，地理上的分离使福古岛的岛民成为手工制作领域的佼佼者，岛民循环利用和修正当地手工制作方式，从而来应对各种挑战。在建筑设计中，所蕴含的文化和智慧的遗产成为一个重要的优势，它们也是建筑的重要资产，这座旅馆为慈善基金会所拥有，由享受经济效益的福古岛社区和Change岛社区运营。

主建筑呈X形布局，二层的东西向体量容纳公共空间，而四层的西南-东北向体量则与海岸线平行，拥有29间客房，旅馆里所有的客房都面朝大海，同时渔场映入眼帘，以吸引游客前往岛屿。

客房的规模从32.5m²到102.2m²不等。三、四层的所有房间内都设有烧木材的壁炉，四楼房间的天花板随着屋顶的斜坡而建，东部的三个房间都是双体量空间，其卧室区位于夹层。公共区域包括福古岛艺术馆创建的艺术长廊、餐厅、酒吧和最近在《Enbroute》杂志排名加拿大酒店前十的休息室；文物图书馆，是已故的私人收藏家Leslie Harris博士的家，Leslie Harris博士是纽芬兰纪念大学的前任校长。

位于二层的电影院同加拿大电影局是合作伙伴关系。四层的屋顶甲板设有桑拿房以及可以欣赏海景的户外热水浴缸。旅馆采用了传统风格的海岸支架来支撑楼层，可同时实现建筑痕迹最小化，并使建筑对围岩、地衣和蓝莓植物的影响降到最低。

自这个项目开始建造以来，生态化的自给自足系统便巧妙地融合其中，通过运用技术来减少和保留水和能源的使用。这座旅馆是一座高度保温的钢框架建筑，窗户都是同等的三窗格玻璃，屋顶的雨水可以被地下室的两个蓄水池收集起来，经过过滤后，用于冲厕和洗衣服，甚至是用于所有厨房内的散热应用。太阳能保温板不仅为地板内的辐射供暖提供热水，也为洗衣房和厨房设备提供了热水。

当地木匠和工匠的知识与技能，对贯穿这座建筑的建筑材料、细节和家具以及纺织品的选用，发挥着至关重要的作用。以他们的技艺为起点，福古岛和Change岛的建造者和手工制作者们与来自北美与欧洲的现代设计师们进行了长期的合作。

Fogo Island Inn

The Fogo Island Inn was conceived by the Shorefast Foundation, a Canadian charitable organization established by Zita Cobb, Anthony Cobb and Alan Cobb, as a building for learnings that have emerged from four centuries of lived experience on the northeast coast of Newfoundland – to help carry the past into the future. It was created to be a cultural and economic engine for Fogo Island, one of Canada's oldest settlements; created in response to a pressing need to find a new relevance for traditional knowledge and traditional ways. The goal was to "find new ways with old things". Fogo Islanders are a people who by virtue of their centuries of geographic isolation have become masters of making things by hand, recycling and devising local solutions to all manner of challenges. Engaging this cultural and intellectual heritage was a key priority in the design of the building and was a key asset in its construction. The Inn is owned by the charitable foundation

1 福古岛的旅馆 2 福古岛的旅馆的附属建筑 3 停车场
4 教堂 5 橙色小屋 6 渔民的大厅 7 墓地
1. Fogo Island Inn 2. Fogo Island Inn's outbuilding 3. parking
4. church 5. orange lodge 6. fisherman's hall 7. graveyard

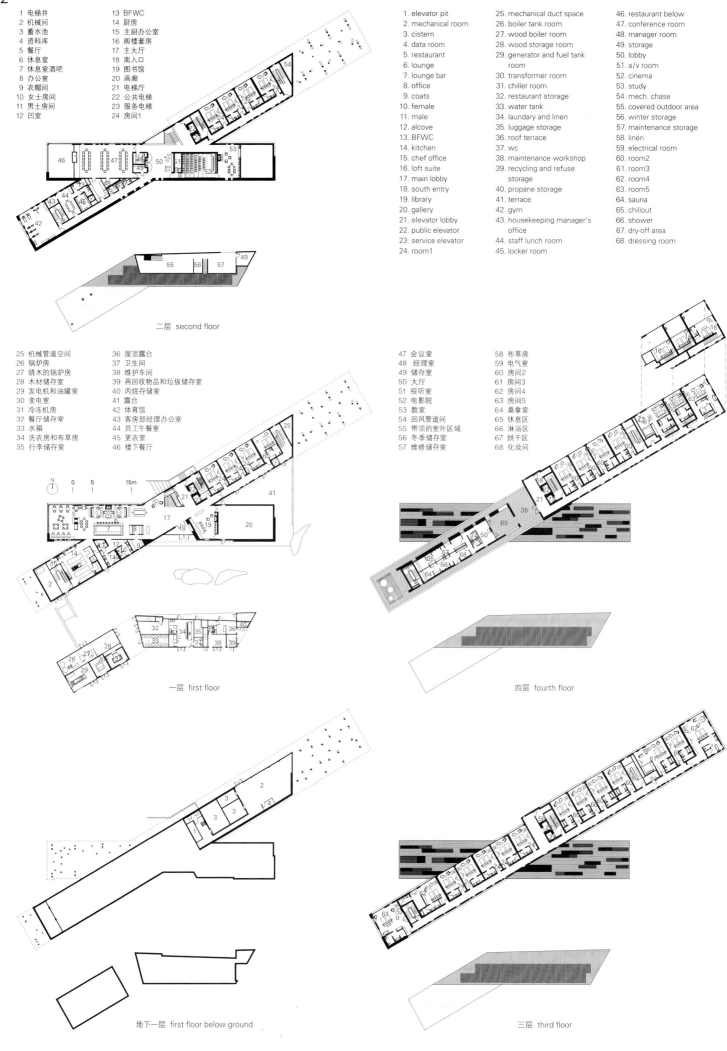

项目名称：Fogo Island Inn
地点：Back Western Shore, Fogo Island, Newfoundland
设计建筑师：Todd Saunders / 委托人：Ryan Jørgensen, Joseph Kellner, Attila Béres, Nick Herder
纪录建筑师：Sheppard Case Architects Inc., St John's, Newfoundland, Jim Case / 主要负责人：Dwayne Gill, Roger Laing
结构工程师：DBA Consulting Engineers Ltd.
机械工程师：Crosbie Engineering Ltd., Sustainable Edge Ltd., Odyssey Mechanical Inc., Bayview Electrical Ltd., Jenkins Power Sheet Metal
建筑管理：Shorefast Foundation / 设计管理：Joseph Kellner, Nick Herder, Kingman Brewster, Eric Ratkowski, Anis Sobhani
景观设计：Shorefast Foundation with consultation from Cornelia Oberlander, James Floyd Associates, Todd Boland and Tim Walsh, MUN Botanical Garden
甲方：Shorefast Foundation
有效楼层面积：4,500m²
设计时间：2006—2010 / 施工时间：2010.6—2013.6
摄影师：©Alex Fradkin (courtesy of the architect) - p.81, p.83
©Iwan Baan (courtesy of the architect) - p.78~79, p.80, p.84, p.85

and is operated for the benefit of the communities of Fogo Island and Change Islands.

The main building is an X in plan with the two-storey west to east volume containing public spaces while the four-storey south-west to north-east volume, parallel to the coast, contains twenty-nine guest rooms. All guest rooms face the ocean and look onto the fishing grounds that attracted people to this island. The room sizes vary from 350 square feet to 1,100 square feet. The rooms on the third and fourth floors all have a wood-burning stove. The ceilings of the rooms on the fourth floor follow the slope of the roof and the three rooms on the east are double volume spaces with the sleeping area located on the mezzanine. Public areas include an art gallery curated by Fogo Island Arts; a dining room, bar and lounge which was recently rated as one of the top ten new restaurants in Canada by *Enroute* magazine; and a heritage library which is home to the private collection of the late Dr. Leslie Harris, the former president of Memorial University of Newfoundland. The second floor includes a cinema that is a partnership with the National Film Board of Canada. The fourth floor's roof deck has saunas and outdoor hot tubs with views of the sea. Traditional-style "shore" legs are used to support the floors while minimizing the overall building footprint and the impact on the adjacent rocks, lichens and berries.

Ecological and self-sustaining systems were subtly integrated from the beginning of the project, incorporating technologies to reduce and conserve energy and water usage. The inn is a highly insulated steel-frame building and the windows have the equivalent rating of triple pane glazing. Rainwater from the roof is collected into two cisterns in the basement, filtered, and used for the toilet and laundry water and also to be used as a heat sink for all of the kitchen appliances. Solar thermal panels supply hot water to the in-floor radiant heating as well as the laundry and kitchen equipment.

The knowledge and skill of local carpenters and craftspeople were essential for establishing the materials, details, furniture and textiles used throughout the buildings. Their know-how was the starting point for what has become a long term and ongoing collaborative project between contemporary designers from North America and Europe and the men and women makers and builders of Fogo Island and Change Islands.

丽阿海滩度假住宅
Mold Architects

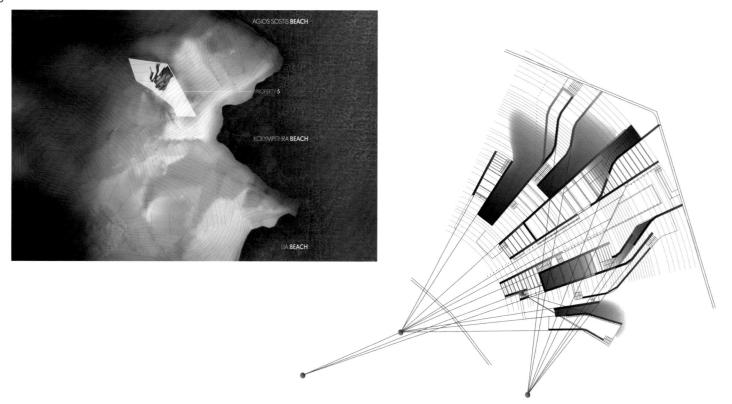

基克拉迪群岛的景观岩石坡由复杂的纵向干石墙组成，其功能是保持那里阶地中的珍贵土壤。

这座建筑设计在一个陡峭的斜坡上，好似一个复杂的"阶地"。这里的干石墙变成了一个组成各个部分的工具，起到了封闭空间、塑造庭院、保护建筑免受北风侵袭和圈定视野的作用，从而创造一个复杂的内外空间布局，自然地融入斜坡的流动线条中。

遮棚结构选择了氧化的H型钢，其更广泛的使用则延续了岛上铁矿开采的历史。利用石头和铁，加上泥土颜色的加固水泥砂浆制成的地面，房子达到了"伪装"的目的。屋顶上种植了可食用的植物，挖出的地方更强化了这种逻辑性，这些连同橄榄树、夹竹桃和附近生长的叶子花属等植物，来对景观进行视觉延续。

房子的设计必须考虑的另一个现实是基克拉迪群岛的严格的建房条款。建筑物的高度不能超过一层，所有的房屋必须要大规模地紧凑合并，不允许使用大孔和悬臂。

岛上的生活很大程度上是在户外发生的，所以建筑师非常重视不同品质的外部空间的创造。带顶的、封闭的、遮蔽的区域必须是完全不同的，而另一些则是没有要求的，能晒到太阳和通风就好。所有的空间都相得益彰，沿路一直延伸到海边。

封闭空间的组成秉承基克拉迪群岛传统的建筑类型，即空间狭小，形状不规则，摆放在一排，还要为以后的扩充留有空间。

设计的目标是把建筑物融入基克拉迪群岛的景观中，最大化地使用方向和景观，在遵从基克拉迪群岛严格的建筑规定要求下，创作一个类型迥异并充分使用外部空间的体系。

为了更好地融入景观，大部分的空间被挖出，在屋顶上种植可食用的植物。同时当地石头成为挡土墙的构成元素之一，其使用也顺应了基克拉迪群岛岩石坡壁上干石墙纵横交错的传统。（鉴于岛上以历史悠久的开采铁矿闻名）铁，连同加固的水泥砂浆、木材和水等也被广泛使用。

Lia Beach Vacation Residence

The rocky slopes of the Cycladic landscape are dominated by the picture of a complex consisting of longitudinal dry stonewalling whose function is to hold the precious soil on the terraces formed there.

The house has been designed as a composite "terrace" on a steep slope. The dry stonewalling here is transformed into a tool of composition which defines the enclosed spaces, shapes the courtyards, gives protection from the northerly winds, and frames the view, thus creating a complex of interior and exterior spaces, in sequence with the natural flow of the slope.

The oxidised IPE beams which were chosen for the construction of the shades, and for more general uses, are a continuation of the

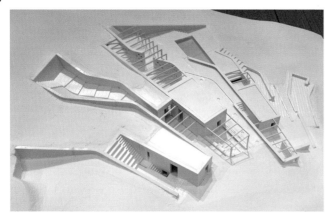

项目名称：Vacation Residence at Lia
地点：Lia Beach, Serifos Island, Cyclades, Greece
建筑师：Iliana Kerestetzi
设计团队：Maria Vrettou, Richard Rubin, Katerina Daskalaki, Fotis Zapantiotis
项目建筑师：Iliana Korootozi
结构工程师：Panayotis Poniridis
结构工程顾问：Andreas Mitsopoulos
机械/电力工程师：Team M-H
室内设计：Iliana Kerestetzi
生物气候设计：LYSIS
能源研究：Team M-H
音效设计：ABAS A.E
景观设计：Iliana Kerestetzi
照明设计：IFI group
承包商：Mold Architects
场地监督：Iliana Kerestetzi
用地面积：6,000m²
总表面积：300m²
设计时间：2011
竣工时间：2013
摄影师：courtesy of the architect-p.88~89, p.93
©Yannis Kontos(courtesy of the architect)-p.86~87, p.90, p.92, p.94~95, p.96, p.97

屋顶 roof

二层 second floor

1 客房和起居室　　1. guest bedroom and living room
2 客用浴室　　　　2. guest bathroom
3 厨房　　　　　　3. kitchen
4 起居室和餐厅　　4. living and dining room

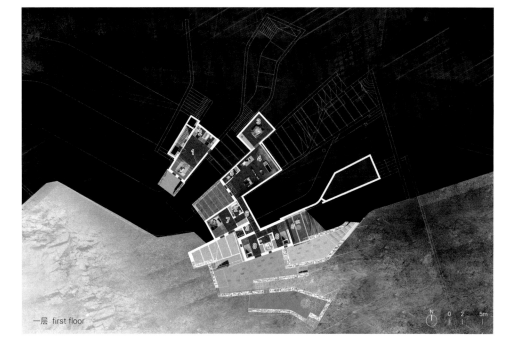
一层 first floor

1 卧室和起居室　　1. bedroom and living room
2 浴室　　　　　　2. bathroom
3 储藏室　　　　　3. storage
4 游乐室/体育馆　 4. playroom / gym
5 化妆室　　　　　5. powder room
6 走廊　　　　　　6. corridor
7 主卧室　　　　　7. master bedroom
8 卧室　　　　　　8. bedroom

island's history(mining of iron ore). By means of the use of stone and iron, in combination with floors of tamped cement mortar in an earthy color, the "disguising" of the house was the aim. This logic is reinforced by the creation of "dug-out" areas, on the roofs where edible plants grow. These, together with the olive-trees, the oleanders, and the bougainvilleas which grow in the surrounding area, contribute to the visual continuation of the landscape.

Another reality to which the design of the house had to respond is the strict terms of building which prevail on the islands of the Cyclades. The height of buildings must not exceed one storey, and the consolidation of all the premises into one compact mass is mandatory, while large apertures and cantilevers are prohibited.

Importance was attached to the creation of differing qualities of exterior spaces, given that life on the island is largely lived in the open air. Roofed, enclosed, and sheltered areas are distinct, whereas others are free, exposed to the sun and the wind. All of them communicate with one another, thus composing the route down to the sea.

The composition of the enclosed spaces was carried out with the typology of traditional Cycladic dwellings, in which spaces of small dimensions, frequently of an irregular shape, are laid out in a row, with a scope for later additions, as a criterion.

The aim of the design is the complete integration of the building into the Cycladic landscape, the best use of the orientation and view, and the creation of a system of differing types and uses of exterior spaces, while at the same time observing the strict building provisions in force in the Cyclades.

For integration into the landscape, most of the spaces are dug-out – edible plants grow on the roofs – while local stone has been used in a composition of retaining walls corresponding to the traditional dry stonewalling which criss-crosses the rocky slopes of the Cyclades. Iron was also used extensively(given that the island is famed for its long history of mining iron ore), together with tamped cement mortar, wood, and water.

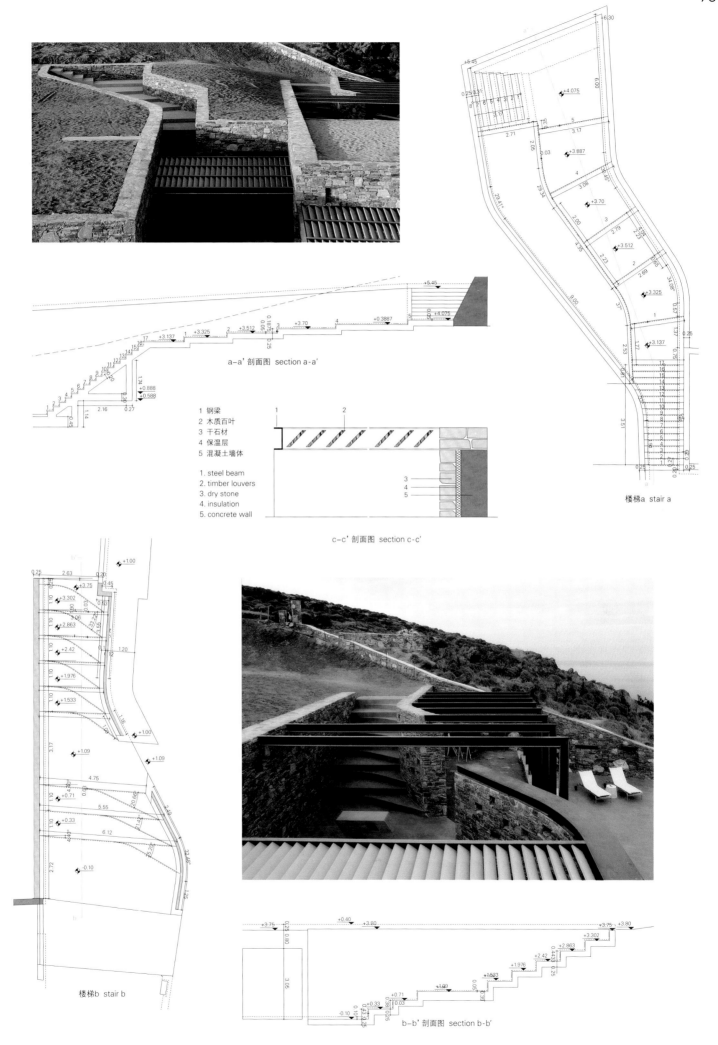

a-a' 剖面图 section a-a'

1. 钢梁
2. 木质百叶
3. 干石材
4. 保温层
5. 混凝土墙体

1. steel beam
2. timber louvers
3. dry stone
4. insulation
5. concrete wall

c-c' 剖面图 section c-c'

楼梯a stair a

楼梯b stair b

b-b' 剖面图 section b-b'

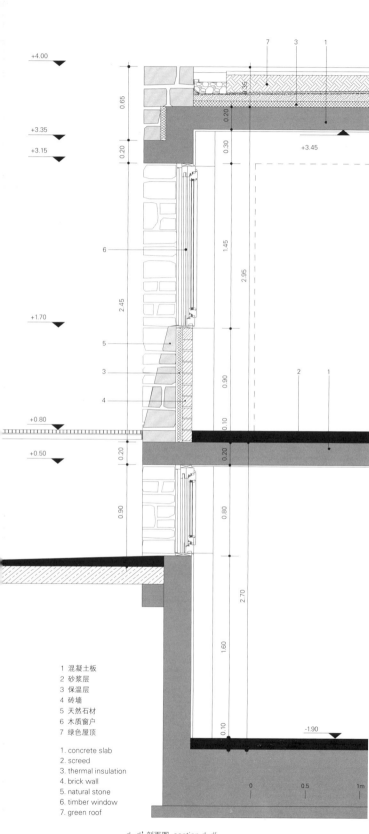

1 混凝土板
2 砂浆层
3 保温层
4 砖墙
5 天然石材
6 木质窗户
7 绿色屋顶

1. concrete slab
2. screed
3. thermal insulation
4. brick wall
5. natural stone
6. timber window
7. green roof

d-d' 剖面图 section d-d'

位于巴库安布兰区,占地58 000m²,安布兰海滩俱乐部以前可容纳1500位客人。但海滩俱乐部现已过时,特别是由于空间不足,在服务方面难以满足其客户的需求。建筑师没有对现有设计进行大幅度的修改,而是采取逐步完善的方法;将集中在海滩和游泳池区的海滩设施扩展到整个房产的每个角落,进而实现对现有设计的重新评估和更新。

该项目寻求在现有的陈旧建筑的周边安置新的功能,并且对立面墙体做出的细微干预和调整,使该建筑物自然地融入到整体建筑的视野中。作为干预工作的一部分,中央水池被保留下来,周围的餐厅建筑也因为增加了新的服务功能而得到升级。

为了增加海滩周围地区和泳池这一狭长房产的主要景点的可视性,深色的露台、酒吧、咖啡馆、休息区平行地修建在房产的两侧。这些设施在设计上略有提高,为整个地区创造了一个低谷效果。

各种娱乐和休闲娱乐区在58 000m²的土地上蔓延开来,人们会设想海滩俱乐部是一个多功能的、各个部分都可投入使用的整体。人行道

安布兰海滩俱乐部

Erginoğlu & Çalişlar Architects

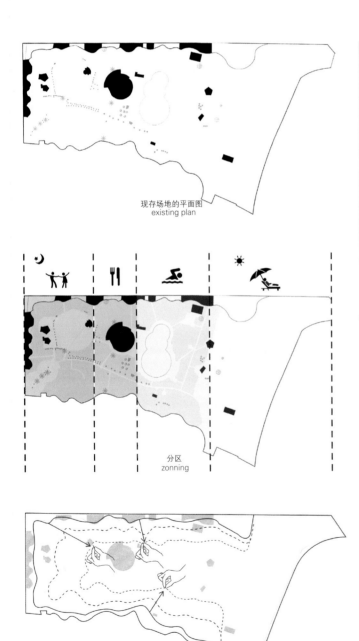

现存场地的平面图
existing plan

分区
zonning

形成天篷
forming the canopy

作为"流动"元素越发重要,它将一系列看似无关的区域连接在一起。婚礼场地、蹦床、植物园、高尔夫场通过横跨整个地区的悬臂梁都连接起来,成为整体的一部分。

说到海滩生活和文化,首先想到的最有名的地方当然是巴西。该项目的目标是将这个美丽的国家的精神传递到不同的地方。该项目借用刚刚去世的奥斯卡·尼迈耶来体现这一精神。甚至沙滩伞的设计方式及视觉上的统一方式都坚持了这种精神。

蜿蜒的走廊和悬臂梁由560根复合柱子与9000m²的钢筋混凝土屋顶组成,体现了建筑的特色并构筑了项目的空间联系。为了使复合材料柱能够支撑14cm厚的钢筋混凝土悬臂梁,建筑师设计了一个特殊的解决方案。为了在地震区建造这样一个悬臂,在柱子连接悬臂的地方设计了锥形开口以降低冲压效果。整个系统不能只依靠钢筋混凝土,而是采用了一个复合系统,把钢筋混凝土注入到厚厚的金属管里。

Amburan Beach Club

Located in the Amburan region of Baku and sprawled across an area of 58 decares, the former Amburan Beach Club catered to a capacity of 1500 guests. No longer able to meet the needs of the present, however, the beach club became outdated and fell short of meeting the needs of its customers, particularly in terms of service, due to spatial shortcomings. Without modifying it, the existing plan has been ameliorated; moving from the idea of spreading the beach amenities centered solely on the beach and pool areas to the entire property, the existing design was thus reevaluated and renovated.

The project sought to assign new functions to the existing obsolete buildings on the circumference of the property; through small interventions and changes made to their facades, the buildings have been made invisible within the overall architectural views. As

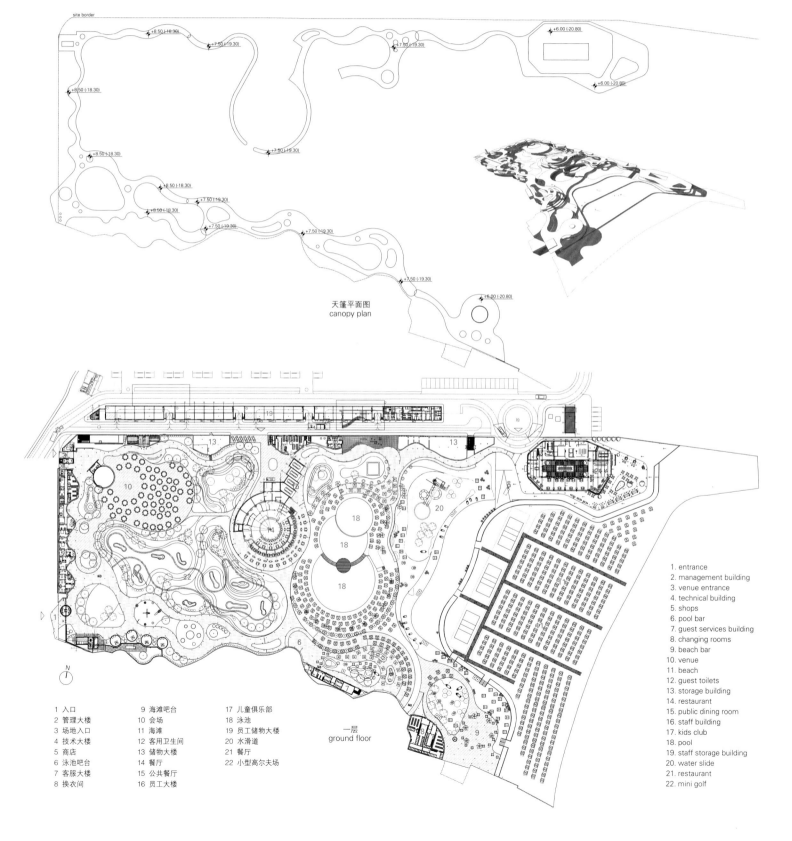

天篷平面图
canopy plan

一层
ground floor

1 入口	9 海滩吧台	17 儿童俱乐部	
2 管理大楼	10 会场	18 泳池	
3 场地入口	11 海滩	19 员工储物大楼	
4 技术大楼	12 客用卫生间	20 水滑道	
5 商店	13 储物大楼	21 餐厅	
6 泳池吧台	14 餐厅	22 小型高尔夫场	
7 客服大楼	15 公共餐厅		
8 换衣间	16 员工大楼		

1. entrance
2. management building
3. venue entrance
4. technical building
5. shops
6. pool bar
7. guest services building
8. changing rooms
9. beach bar
10. venue
11. beach
12. guest toilets
13. storage building
14. restaurant
15. public dining room
16. staff building
17. kids club
18. pool
19. staff storage building
20. water slide
21. restaurant
22. mini golf

part of this intervention, the central pool has been preserved and the restaurant building around it has been upgraded through the addition of new service units.

In order to lend more visibility to the areas around the beach and the pool, which constitute the main points of attraction upon the long and narrow property, the tanning terraces, bar and cafes, and relaxation/lounging areas have been set parallel along the two long sides of the property. They are slightly elevated in order to create a valley effect across the entire area.

Hence, the various entertainment and recreation areas are spread homogenously across the 58-decare land and the beach club is thus conceived as a multi-functional whole, each part of which can be put to use. The walkways gain significance as a "flowing" element that connects and surrounds a range of seemingly disconnected areas upon the property. Various functions such as a the wedding venue, trampoline, botanical gardens, and golf course are thereby associated with one another as part of a whole through the use of a cantilever that runs over the entire area.

In speaking of beach life and culture, the first and most famous name to spring to mind is, of course, Brasil. The project aims to carry the spirit of this beautiful country to an entirely different geography. In doing so, the project strives to reflect this spirit by a reference to Oscar Niemeyer, who passed away recently. Even the beach umbrellas are designed in a manner and visual unity that uphold this spirit.

Constituting the architectural character and spatial connections of the project, the meandering walkway and cantilevers are comprised of 560 composite pillars and a nine-thousand-square-meter reinforced concrete roof. For the composite pillars to carry the 14 cm-thin reinforced concrete cantilever, a special solution has been designed uniquely for this project. In order to create such a cantilever in an earthquake zone, conical openings have been made at the points where the pillars meet the cantilever to reduce the punching effect. As this system cannot solely be created by using reinforced concrete, a composite system has been employed by injecting reinforced concrete inside a thick sheet metal pipe.

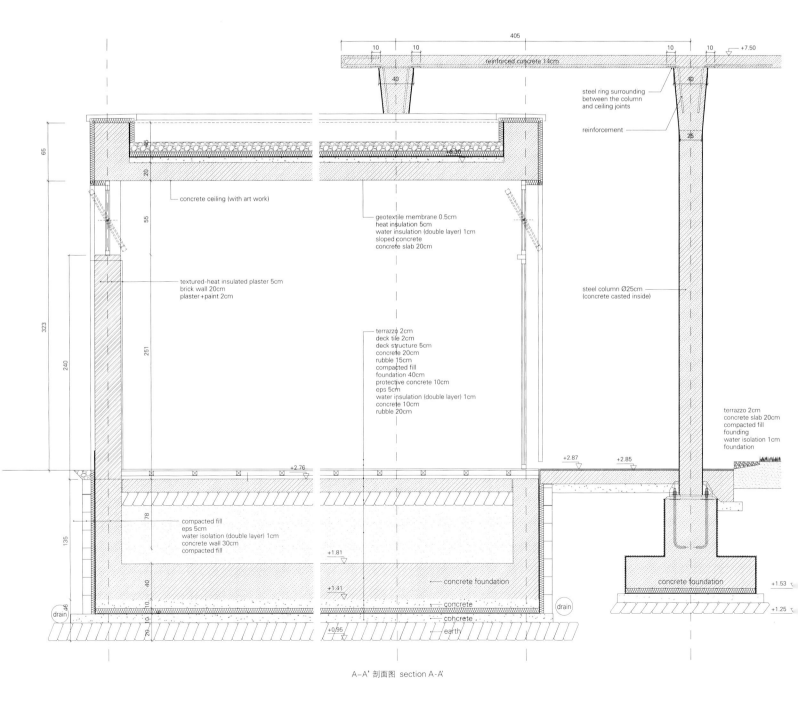

A–A' 剖面图 section A-A'

项目名称：Amburan Beach Club
地点：Bilgah District, az1212 Bilgeh Baku, Azerbeycan
项目团队：İ. Kerem Erginoğlu, Hasan C. Çalıslar, Fatih Kariptas, Fulya Arabacıoğlu, Serhat Özkan, Arman Akdoğan, Ezgi Sönmez, Felix Madrazo, Bas van der Horst, Füsun Seçer Kariptas
机械工程师：Arkon
电气工程师：Enkom, Teknoel
景观建筑师：Arzu Nuhoğlu
承包商：Pasha Construction
甲方：Pasha Construction
用地面积：11,000m² 总建筑面积：4,200m²
施工系统：composite system
结构设计：Modern Mühendislik
竣工时间：2013
摄影师：©Cemal Emden (courtesy of the architect)

维拉多姆斯儿童夏令营

OAB – Office of Architecture in Barcelona

新的营地将包括三个不同的区域:服务楼、自然课程教室和休息区。服务楼里有餐厅、厨房、储藏室/仓库,以及接待区/信息区。

餐厅可容纳100位宾客,在这个即将建造的大房间里,留在夏令营中的不同的群体可以发现一处会议空间和社交区域。餐厅隔壁的建筑将包含三间自然课程教室,每间能容下30人。有别于传统度假屋的"梳子式格局"(长长的走廊两侧是很多并排的梳子齿式的房间,公用的洗手间在大厅的中间或一侧),建筑师要建造私密却又互相联系的,而且自主的空间单元。建筑的容积率增大,外面的通道使单元之间的沟通交流变得可能且畅通。这种安排让不同群体可以生活在同一个设施条件下,但在各自的房间内仍保持一定程度的隐私。

体量的分割设计允许不同的小型自主单元对用户的数量进行划分,并且使维修和管理的费用降到最低,同时使整个系统的布局还是一个整体。

建筑单元被规划成一个基准原型,描绘了一幅儿童世界的景象:微型房间,微型人群,与森林、道路和自然的关系(使系统本身能够轻易地拓展)。

建筑师设计了三种房间:四人儿童屋、六人儿童屋和八人儿童屋,主体部分可以分成一个或两个层次,实现了总体量多达90个地方的目标。因为房间融合了声控系统和人工照明系统,且天花板上安装了垂吊的白色灯管,所以上部空间和空气含量因此变得非常重要。

由于使用简单的材料,因此在实施、最终产生的影响和日后的维修方面会带来更大的经济意义。在室内,颗粒精细的混凝土砌块做成的地面显露出来,并且进行了涂漆和抛光,20cm宽的窗户嵌入到2mm厚、带有百叶的哑光不锈钢架内,将室内变暗,结合榫槽接合的圆顶,给室内光带来温暖的色调。垂直窗口的位置使气流在无论刮哪种盛行风的情况下都通风良好。

室外的立面和屋顶使用了特殊材料，无论是各自的房间还是公共的空间，都对Coteterm吸附分离系统做出了响应——即使用了灵活的自洁性灰泥。这既体现了连续性，也使建筑物的四周可以防水绝缘。室外道路被分为两种类型：供行人行走的带有纹理的混凝土板和粗糙的混凝土石块。

Viladoms Children's Summer Camps

The new camp would consist of three different zones: the service building, the nature classrooms, and the sleeping area. The service building is where one can find the dining room, the kitchen, the pantry/warehouse, and a reception/information area.

The dining room will have a capacity for about 100 guests, and will be created as just one large room where different groups that stay at the camp could find a meeting space and social area. The building next door to the dining room will contain the three nature classrooms that will each have a capacity for about 30 people. Breaking off from the conventional types of holiday homes which only follow the "comb scheme" – one long hallway with lots of rooms running adjacent as though tines in a comb, and communal bathrooms in the middle or end of the halls – the architects propose to create individual units that will all relate to each other, but still remain autonomous units. The ratio of floor area increases, and communication between the units is possible through outer walkways. This modular arrangement allows, for example, accommodation for different groups within the same facility, but

项目名称：Viladoms Children's Summer Camps
地点：Road BV-1212 Km.5,15 – Castellbell i el Vilar
建筑师：OAB – Office of Architecture in Barcelona
项目团队：Carlos Ferrater, Nuria Ayala
合作建筑师：Alexandre Pararols
甲方：Fundació Catalana de L'esplai
城市化区域面积：1,449m²
有效楼层面积：770.48m²
造价：EUR 504,079
设计时间：2010
施工时间：2010—2011
摄影师：©Alejo Bague (courtesy of the architect)

1 餐厅　2 教室　3 旅馆1　4 旅馆2
1. dining room　2. classroom　3. hostel-1　4. hostel-2
一层　first floor

the possibility of still maintaining some degree of privacy in the "houses."

Volumetric fragmentation allows small autonomous units to graduate the number of users and minimize the costs of maintenance and supervision, while allowing configuration of a system within the whole colony.

The unit is proposed as a benchmark archetype and image of the imaginary world of childhood: small houses, small people, relationship with the forest, roads, and nature that allows for easy extension and expansion of the system itself.

The architects propose 3 different room types, with groups of 4, 6 or 8 children, with the opportunity to develop a level or two as the body, resulting in the end of a total capacity of 90 places. The upper space and volume of air have acquired great importance given that the ambience of the rooms favors the incorporation of acoustic control and artificial illumination provided by white, tube lighting suspended from the ceiling.

The simplicity of the materials used has allowed greater economic viability by means of implementation, final impact, and subsequent maintenance. In the interiors, exposed, painted, and polished concrete block floors of fine particle size, window width of 20cm embedded in matte stainless steel frames of 2mm thick with painted shutters, incorporated to darken the interior, and tongue and groove jointed domes all come together to give a warm tone to the interior light. The location of the sets of vertical windows allows perfect cross ventilation, regardless of the prevailing wind in each case.

For the exterior, the uniqueness in the materials of the facades and roofs, both in the cellular rooms and public buildings, responds to a system of Coteterm de Parex – a flexible and self-cleaning stucco that allows for continuity and insulation around the water-repellent treated perimeter. The outdoor pathways are differentiated into two flooring types: textured concrete planks for pedestrians, and rough concrete boulders.

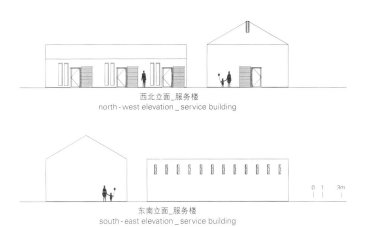

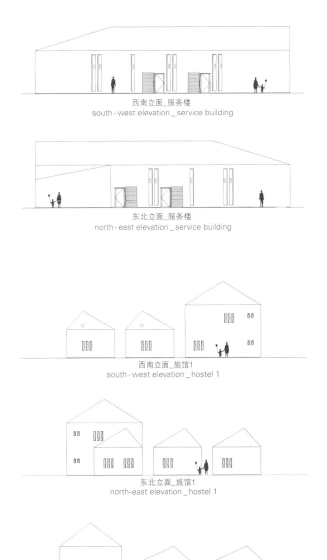

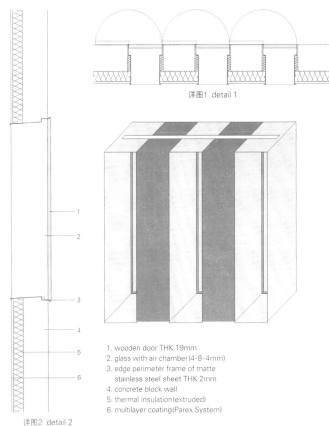

– 1x1m window which consists of 3 modules
– frames made of stainless steel sheet THK 2mm
– covered by thick wooden door

详图1 detail 1

1. wooden door THK 19mm
2. glass with air chamber (4-8-4mm)
3. edge perimeter frame of matte stainless steel sheet THK 2mm
4. concrete block wall
5. thermal insulation (extruded)
6. multilayer coating (Parex System)

详图2 detail 2

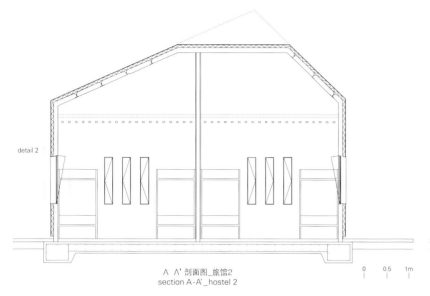

A-A' 剖面图_旅馆2
section A-A'_hostel 2

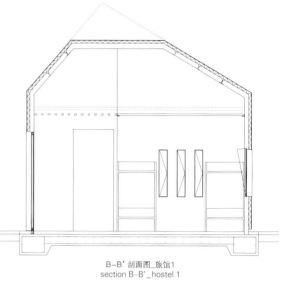

B-B' 剖面图_旅馆1
section B-B'_hostel 1

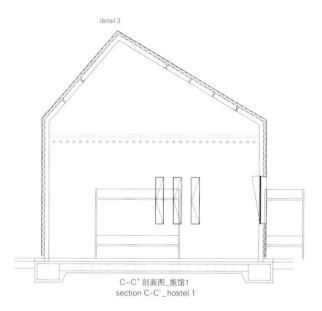

C-C' 剖面图_旅馆1
section C-C'_hostel 1

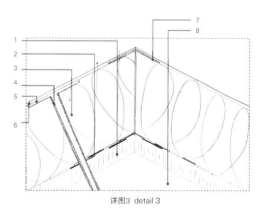

详图3 detail 3

1. Lanko 221
   Coteterm multilayer system of Parex
2. waterproof insulation
3. thermal insulation
4. Lanko 603
   Coteterm multilayer system of Parex
5. Cotespiga E-90
   Coteterm multilayer system of Parex
6. Coteterm mesh(2 layers)
   Coteterm multilayer system of Parex
   flexible stucco
   Coteterm multilayer system of Parex
7. recoil ridge
8. concrete

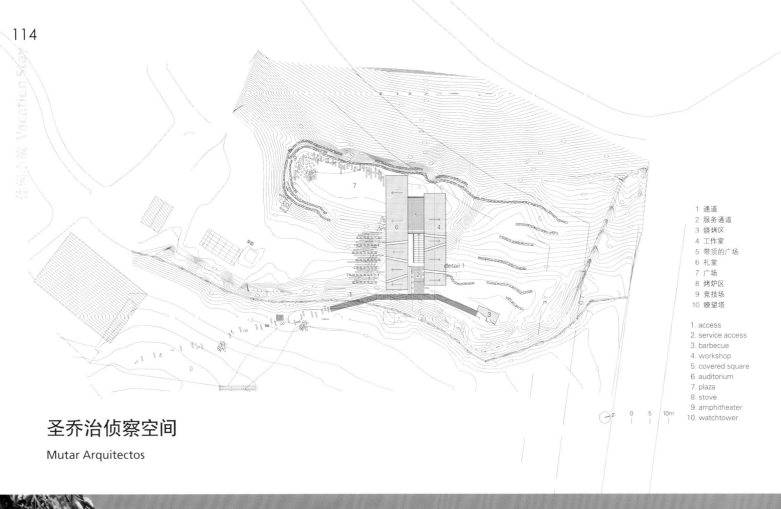

1 通道	1. access
2 服务通道	2. service access
3 烧烤区	3. barbecue
4 工作室	4. workshop
5 带顶的广场	5. covered square
6 礼堂	6. auditorium
7 广场	7. plaza
8 烤炉区	8. stove
9 竞技场	9. amphitheater
10 瞭望塔	10. watchtower

圣乔治侦察空间
Mutar Arquitectos

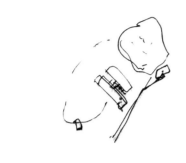

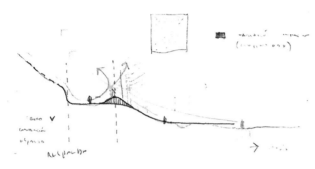

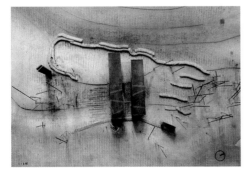

本项目和另外14个提案是在圣乔治大学组织的一次竞赛中提出的。比赛的目的是为学校设计新的侦察空间，它包括2个多功能房间，5个用于存储不同童军用品的地下室，可容纳300人的用于烧烤的户外广场，以及带顶的可容纳150人的广场。

学校把地点定在玛奇哈山的斜坡上，四周榆树环绕，附近有一个废弃的池塘，还有一条水渠流过学校及周边地区。

为了融入当地的主要多元性要素，建筑师采用了分散而非集中的方法。整个项目因地制宜，就地取材，定制化的功能置于池塘上、斜坡上、榆树林里和参观路线的沿岸，成为主要的景观。

修剪了灌木和黑莓植物，砍除了快要倒塌的树木，去除了池塘东部的边缘，这些简单的人为干预使景致的自然性得到平衡。山坡、学校、球场、城市被有机地连接在一起。

这个项目与纵向的、包含3个空间元素（池塘/山坡/水渠）的、双体量的主功能区遥相呼应，每个体量得以在双倍高度下俯瞰学校，并都可通过与池塘处在同一水平线上的户外广场进入其内。

于中间的位置，体量和室外广场的中间建起了带顶的广场，它容纳了多功能房间，具有功能外显化的特点，使整个规划特别是表面显示了更大的通用性。两个体量在底部通过外面游廊连接在一起，游廊成为为地下室提供服务的通道，同时通过在废弃的铁路枕木上修建的宽阔的楼梯与上面相连。

建筑附加了一个木质平台，贯穿整个项目，与水渠平行，烧烤区（榆树林）位于其上，周围满是来自森林的石笼。这里成为人们在户外广场聚集的平台。

场地使用的贵重材料已达到和谐统一，如木材和混凝土，中性的染料凸显了自然的颜色。

黑色木材的外部和切割后露出的白色内部形成巨大反差，突显了外立面，犹如光影斑驳，寻找光明的森林。

位于户外广场边界上的瞭望塔，从不同层次上提供了一种视觉照应。学校强调安全性，并承诺要为侦察活动做出重要贡献。

Saint George Scout Space

The project was presented with another 14 proposals in a contest organized by the Saint George's College; the purpose of the contest was to design the new scout place for the school, which had the following program: 2 multi-purpose rooms, 5 cellars used for storing the different scout groups goods, an outdoor square for 300 people's barbecue and a roofed square for 150 people.

The ground location was defined by the school in the slopes of the Manquehue Hill, rounded by a forest of elms, a disused pond and a water channel that runs through the school and other neighboring sites.

For the incorporation of the main elements of the place, the architects fragment the program, avoiding concentration. With this decision, the project takes hold of the place in extent, placing customized program over the pond, the slope of the hill, the forest of elms and along the travel channel, dominating the views of the landscape.

项目名称：Scout Space, Saint George College
地点：Via Morada n° 5400, Vitacura, Santiago, Chile
建筑师：Claudio Molina Camacho, Daniel de la Vega Pamparana, Eduardo Villalobos Fornet
项目经理：Juan Silva
结构工程师：Alvaro Velez
施工：Alzerreca y Diaz Ltda
施工监理和技术员：Jorge Cifuentes
捐助者：Jorge O'ryan
甲方：Saint George's College
用地面积：3,500m² / 总建筑面积：400m²
材料：wood, concrete, steel, stone gabion
设计时间：2008 / 竣工时间：2009
摄影师：©Sergio Pirrone

To avoid unnecessary processing of the natural context is proposed as interventions in the field, only a general thinning of shrubs and blackberries, the cutting of trees in danger of collapse and removal of the eastern edge of the pond, resume the continuity of the hill, the school, its courts and the city.

The project responds to the main program with two volumes that longitudinally inhabit 3 instances of the place (pond/slope of the hill/channel). Thus each of these volumes overlooks with double height toward the school and is accessible from the outdoor square in the level of the pond.

It is in the intermediate space, between volumes and outdoor square the roofed square is generated, incorporating multi-purpose rooms, externalized, giving greater versatility to the program and surface. Both volumes are unified in its lowest level through an exterior covered corridor which defines the service access for the cellars, having connection with the upper level through a broad staircase stands on disused railway sleepers.

To this is added a wooden platform that runs the program, through its parallel to the channel access with the barbecue topping(forest of elms) and gabions of stone from the woods across the project to generate the stands for the mass meeting in the main outdoor square.

Seeking harmony with the place-defined noble materials, like wood and concrete, the project is treated with neutral dyes to highlight the colors of nature.

The contrast between the black wood exterior and the white wood cuts interior highlights the facades, as an analogy of the forest in search of light, defined by contrasts of light and shadow. The watchtower is located at the perimeter of the main outdoor square, serving as a visual reference at different levels of approximation. The college strengthens security and compliance program as an important contribution for scouting activities.

在森林中寻找光明
seeking the light in the forest

追踪路径
tracking the path

树干
trunks of trees

底片中的树干
trunks in negative

贯穿森林的路径
path through the forest

追随水的声音
accompanying with the sound of water

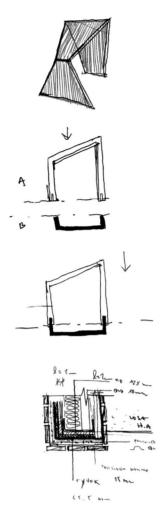

南立面 south elevation

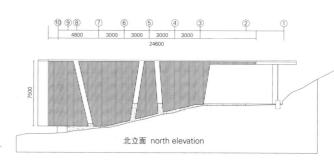

北立面 north elevation

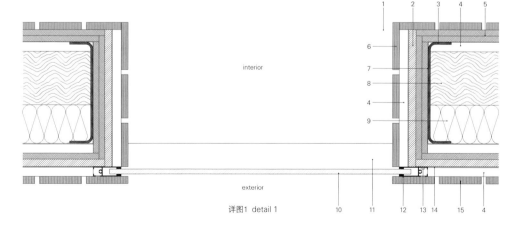

详图1 detail 1

1. reinforced concrete slab/plastered/vitrified, 130mm
2. structural plywood, 15mm
3. FE channel section, 200/50/3
4. metal con 40r support, 40×18×10, e=0.5mm
5. volcanite drywall,
   interior=RF 12.5mm / exterior=rRF 12.5mm
6. vertical batten marine plywood, 2400×90×15 @10mm
   treated with white latex diluted at 50%
   wetproof waterproofing treatment
7. polyethylene, 0.03mm
8. planed pine beam cut to size, 2"×8"
9. mineral wool, 11kgm/m³, 80mm
10. polycarbonate, 10mm
11. FE angle section, 50/50/3
12. 2015 high baseboard section-crown aluminium,
    1.5mm thick
13. fixing screw, 8mm
14. felt 15 lbs
15. vertical batten marine plywood 2400×90×15 @10 mm
    treated with semi-transparent weatherproof chilcoStain
    (no.15 black) on every side, diluted to 75%

1 带顶的广场
2 服务通道
3 室外带顶的游廊
4 通道
5 工作室
6 衣帽间
7 起居室
8 礼堂

1. covered plaza
2. service access
3. outdoor covered corridor
4. access
5. workshop
6. locker room
7. living room
8. auditorium

A-A' 剖面图 section A-A'

B-B' 剖面图 section B-B'

C-C' 剖面图 section C-C'

D-D' 剖面图 section D-D'

朝圣旅馆

Sergio Rojo

一处场所如果长期被用于临时性用途,往往就会疏于管理维护,当使用者事先得知该场所可使用的时间非常有限时,就会渐渐失去对它的关注。

这就是发生在老艺术学院建筑的墙上的事实,它们在几十年暂时性的使用中幸存了下来,这里的居住者都来自于不同的环境,将自己在这里长久的生活想像成了短暂的居留。因此这座建筑在20世纪的下半叶遭受了严重的退化。其建筑师可能是Jacinto Arregui,他也是慈善大厦(1864年)和省立医院(1866年)的设计师。

起初,该建筑临近省立医院(其前身是一个喜剧剧院),导致其后期又被一座新剧院占用了。再后来,其规模和独特的表演空间(高度高,跨度长而且没有柱子和包厢)正适合作为施粥场的餐厅。然后,临近Pastrana建筑这样的条件(它的大门相对,经由市长街可以直接到达学院)被殡仪馆所有者看做是一个买入的好机会。在20世纪的最后几年间,伴随着该地区在时间的长河里逐渐衰退,这里变成了一个废料场和车库。

对于文化遗产缺乏鉴赏力已不是什么新鲜事。其根深蒂固的程度远超人们的想象。需要强调的是,不仅是一系列的历史兴衰,还有当今的环境,都让今天的圣地亚哥使徒庇护所能够促进城市与周围环境的紧密连接,对周围城市肌理产生直接的影响。

建筑师将建筑和城市肌理这两种概念理解为一个不可分割的二项式。如果要提倡单一建筑物的复原工作,则建筑师不应该摒弃或遗失容纳它的公共空间的一致性。

如今,幸运的是,这个故事在这座旧建筑身上得到了逆转。新的所有者明智地衡量了它在建筑学上的潜力和近几年来濒临消失的现状,其历史背景和建筑实体藉此都被开发利用了起来。

结合这个二项式的第二个组成——城市肌理,凭借此时此刻这条路所具有的纪念碑作用,建筑师最终完成了这项再生和修复工作,因为这个任务是必要的,且超越了建筑本身。

因为道路会扩展城市的连接网络,因此,首先,建筑师要将住所与其他詹姆斯一世时期的里程碑,诸如皇宫教堂(城市中的第一家朝圣医院)、石桥和圣格雷戈里奥教堂联系到了一起。这一点得到了肯定与巩固,第二,它目前仍在使用中,这样一定加速了房客的随机流动率,使它的维护工作变得艰难。

Pilgrim Hostel

The temporality of a place that is occupied provisionally, often precipitates the lack of interest in its maintenance, and when it is known beforehand that the use of a space will be limited in time, the concern for it is undermined.

This is what happened to the walls of the old Liceo Artístico, which survived for decades to temporariness, as its inhabitants, all from different conditions, conceived their permanence in it as something ephemeral. And hence the significant deterioration suffered during the second half of the twentieth century. Its architect may have been Jacinto Arregui, the author of the Charity Building(1864) and the Provincial Hospital(1866).

At first it was its proximity to the Provincial Hospital (where its predecessor was the comedy theater), which led to its occupation as a new theater. Later, its size and unique performance space (great

之前的北立面
previous north elevation

现在的北立面
current north elevation

height, long spans without columns, boxes) were suitable to install the dining room of the soup kitchen. Then, the proximity to the Pastrana Building (its doors were facing, as the Liceo was accessed through Mayor Street) was seen as a great buying opportunity for the owners of the funeral home. During the last years of the twentieth century, and accompanying the decline over time in the area, it became a scrap yard and garage.

The lack of appreciation of our heritage is not something new. It is more ingrained in us than we think. People must emphasize that not only this peculiar succession of historical vicissitudes, but its situation, enabled today's Santiago Apostle Shelter to catalyze an intense network of city links with its near surroundings, and to have its direct impact on the urban fabric of the area.

The architects understand these two concepts, architecture and urban fabric – as an indissoluble binomial. And if the architects promote the rehabilitation of unique buildings, they should not spurn or lose the uniqueness of the public space containing them. In this old building, now, fortunately, the story is written backwards. The new property has intelligently weighted the potential of its architecture, endangered in recent years, and has used it as leverage to boost its exploitation in a sensible way with its history and its entity.

Combining the second component of the binomial, the urban fabric, through the monumental force that the road has at this point, the architects end up completing a regenerative and rehabilitating action, because it is necessary, and beyond the building itself.

Because the road will expand the network of urban links, firstly, the architects should link the shelter with other Jacobean milestones, such as the Imperial Palace Church (First Hospital of pilgrims in the city), the stone bridge and the chapel of San Gregorio. It nods and consolidates, secondly, its continues the use in time, thus definitely turning away a random rotation of tenants that made its preservation difficult.

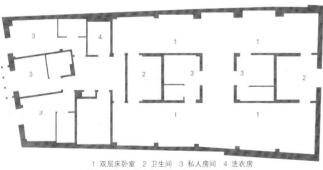

1 双层床卧室 2 卫生间 3 私人房间 4 洗衣房
1. bunk beds 2. toilets 3. private room 4. laundry
二层 second floor

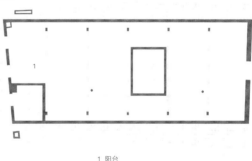

1 阳台
1. balcony
屋顶下层 below roof floor

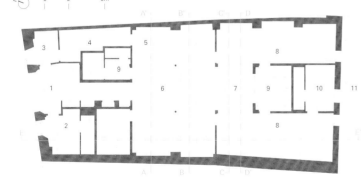

1 入口 2 主人房间 3 仓库 4 厨房 5 酒吧 6 早餐室 7 露台/小径
8 双层床卧室 9 卫生间 10 私人房间 11 后院
1. entry 2. host room 3. warehouse 4. kitchen 5. bar 6. breakfast room
7. patio/foot bath 8. bunk beds 9. toilets 10. private room 11. back yard
一层 first floor

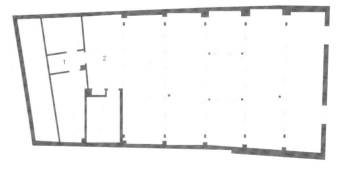

1 建筑系统 2 阳台
1. building systems 2. balcony
三层 third floor

128

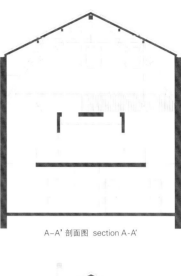

A-A' 剖面图 section A-A'

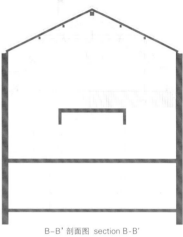

B-B' 剖面图 section B-B'

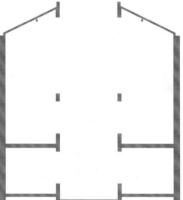

C-C' 剖面图 section C-C'

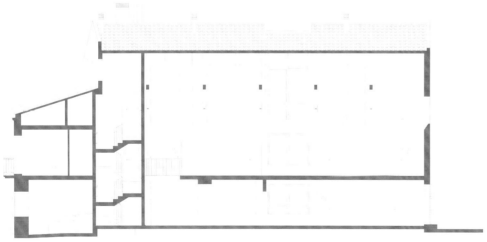

E-E' 剖面图 section E-E'

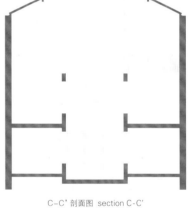

D-D' 剖面图 section D-D'

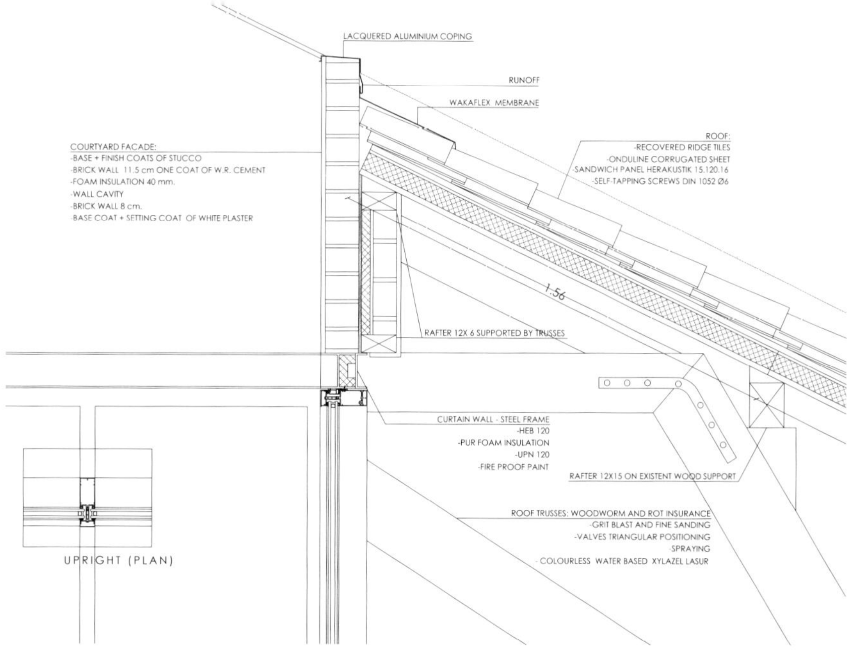

详图1 detail 1

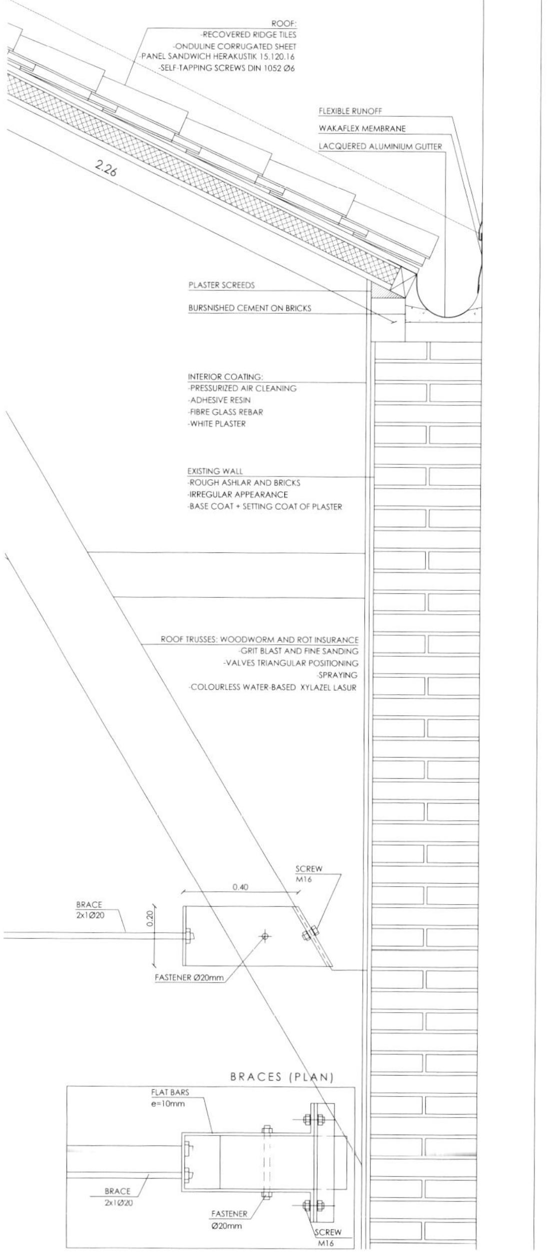

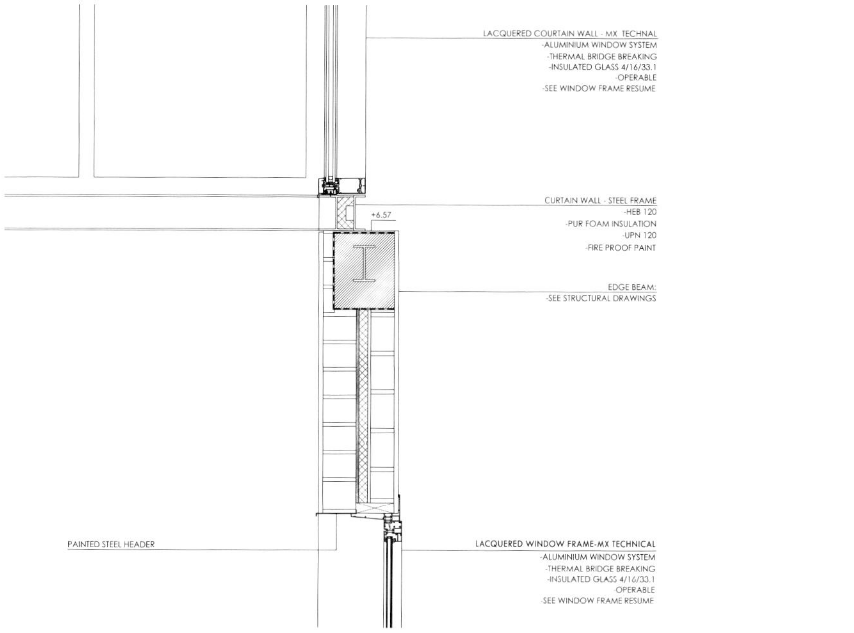

详图3 detail 3

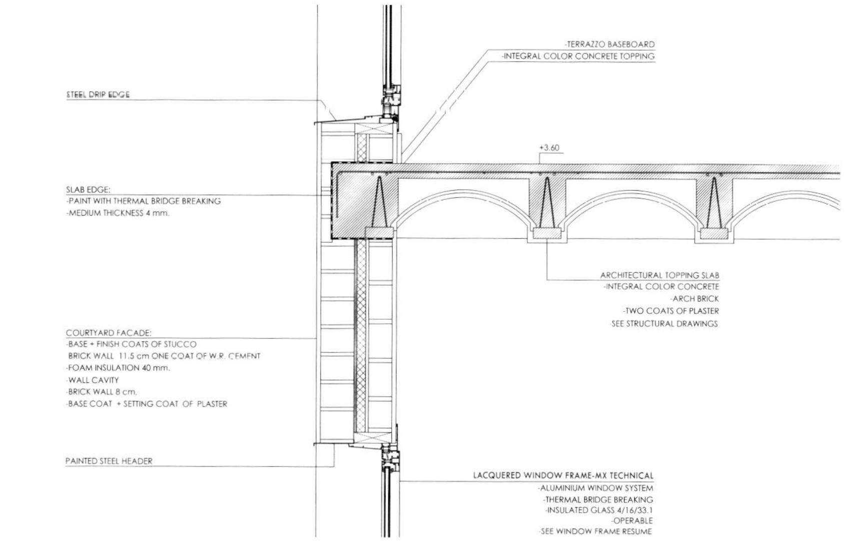

详图4 detail 4

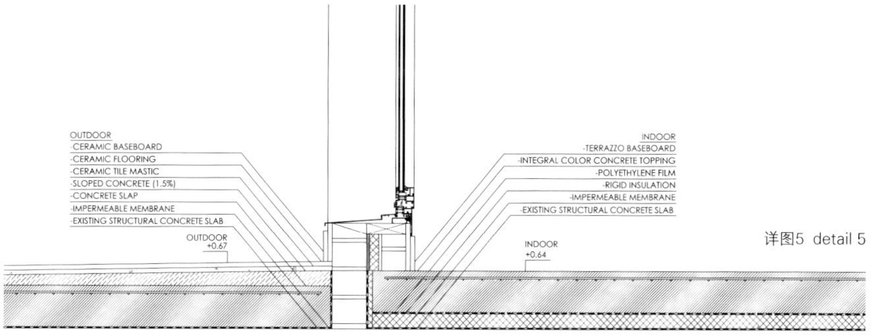

详图5 detail 5

详图2 detail 2

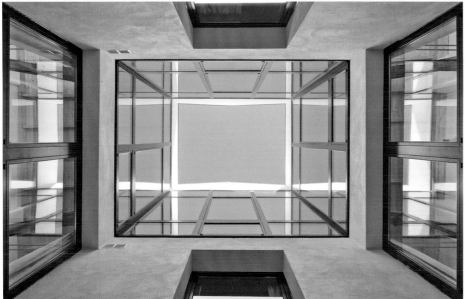

项目名称：Pilgrim Hostel
地点：Ruavieja 42, Logroño, La Rioja, España. Postal Code 26001
建筑师：Sergio Rojo
考古学家：Carlos López de Calle, Juan Manuel Tudanca
工料测量师：Javier Martínez Pérez
结构顾问：Bessel. José Luis Gutierrez, Mercedes Blanco
建筑系统：De Orte Consultancy – Javier De Orte
用地面积：362m² / 总建筑面积：305m² / 有效楼层面积：650m²
设计时间：2012—2013
竣工时间：2013.11
摄影师：©José Manuel Cutillas (courtesy of the architect)

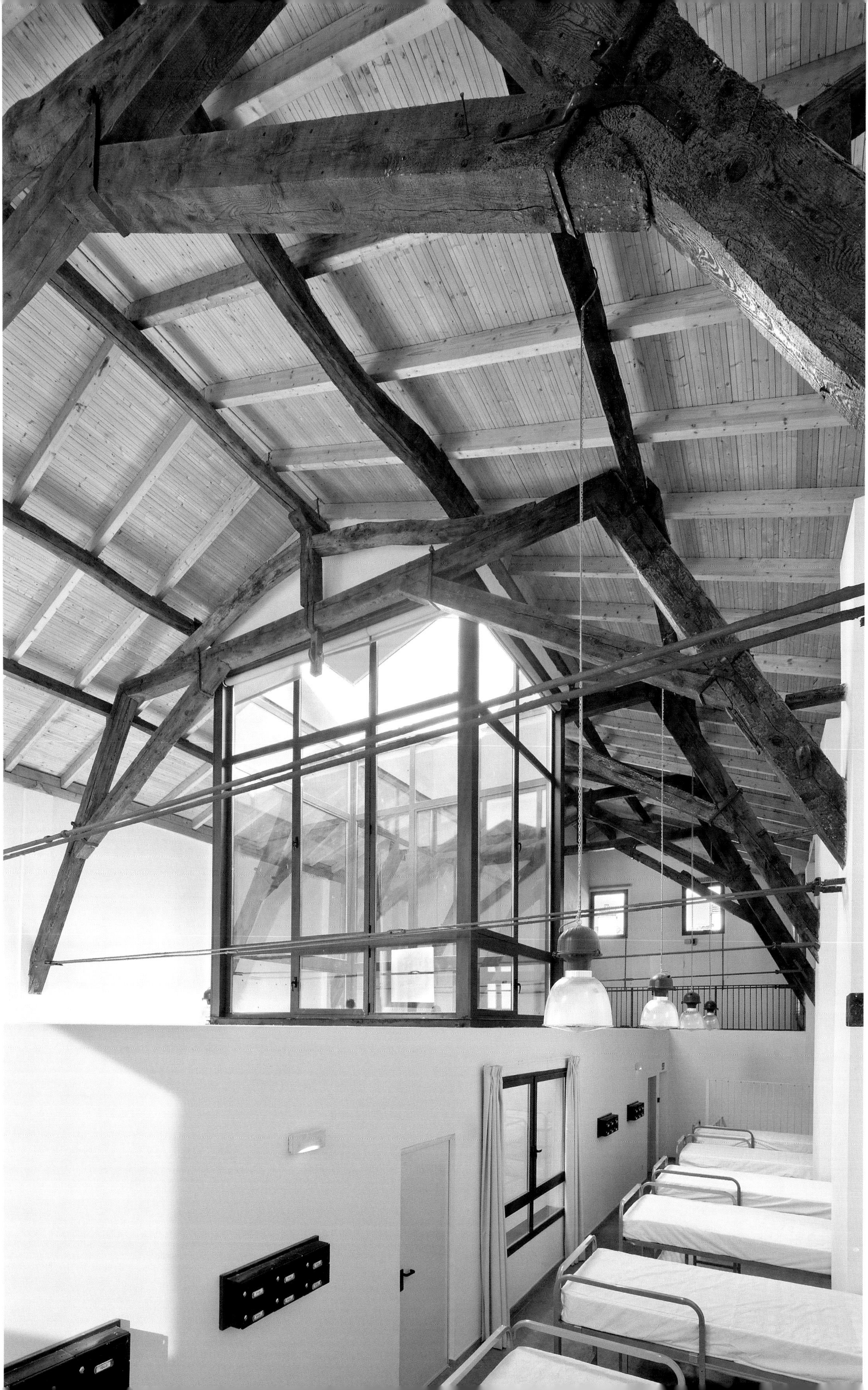

住宅与社区
Dwelling and Comm

在过去的一个世纪里，有关私人同公共领域的关系的观点不断地发生着改变，并已经主要影响到房屋概念的发展。然而房屋单元主要是为家庭或者个人营造的空间，在时间的长河里，住宅同社区间的交流以不同的形式发展着。

住宅体块聚集形成的庭院，驻扎在20世纪转折时期的柏林和阿姆斯特丹此类城市的肌理中，成为一处重要的缓冲区域，用来协调私人同集体领域间的关系。自由体量的新范例在内战时期受到了勒·柯布西耶和国际现代建筑协会的《雅典宪章》的拥护，并且致力于修正社区作为通过开放的绿色区域来相互连接的邻里单元这一理念。自20世纪50年代起，建筑设计和理论的政治层面寻求一个更具人性化的方法来设计居住区，以进一步加强私人和公共领域之间的联系。这个过程导致的尤其引人注意的一个结果是引导人们更加亲近大地，也直接导致高密度的低层住宅区的出现。

现在，在上个世纪里融入建筑行业金科玉律的房屋理念如今仍然在世界范围内取得了不同程度的成功。因此，我们可以观察到，在挖掘私人和公众领域之间的界限的模糊定义的潜能方面，这一行为趋势呈上升状态。因此，提高建筑的多孔性作为建筑操作中最常用的方法之一，与协调个人与公共之间的关系相关。在这种环境下，增加的孔隙工作将作为一个（抽象的）工具，来激活住宅与城市之间千丝万缕的关系。

Over the last century, the changing views on the relation between the private and the public realm have chiefly influenced the evolution of housing concepts. While the housing unit has always been primarily a space for the family and for the individual, the interfaces between the dwelling and the community evolved through time in different ways.
The collective courtyard of the housing blocks that populated the urban fabric of cities such as Berlin or Amsterdam at the turn of the twentieth century, performed as an important buffer zone to negotiate the private with the collective realm. The new paradigm of the freestanding block championed in the interwar period by Le Corbusier and CIAM's *Athens Charter* contributed to revise the idea of community as a neighbourhood unit articulated through an open green space. From the 1950s on, the politics of architectural design and theory sought a more humanistic approach to the design of the habitat, which should develop a stronger connection between the private and the public realms. One of the most noticeable consequences of this process was a drive to bring people closer to the ground, a process that resulted in the emergence of a trend to develop low-rise housing with high density.
Nowadays, the housing concepts that pervaded the architecture discipline over the last century are still performing with different degrees of success throughout the world. One can observe, nevertheless, there is a growing tendency to explore the potential of an ambiguous definition of the boundary between the private and the public realm. Hence, increasing the building's porosity stands out as one of the most frequent tropes of architectural operations engaged in reconciling the individual with the public. In this context, increasing porosity works as a device to activate the entwined relation between the *domus* and the *polis*.

Enhancing Porosity: Rethinking the Interface Between the Dwelling and the Community

The House and the Street
Some of the urban models that influenced the development of urban design through the two centuries surfaced as an attempt to accommodate and control the masses. In Europe, the perimeter block received widespread acceptance, from the late nineteenth century through the first half of the twentieth century, for it became an optimal solution to cope with the rapid development of industrial regions and to control the social problems associated with the growth of the population. One of the main facets of this model was the straight definition of boundaries between the public and the private realm. The perimeter block defined a sharp transition between the street, the dwelling, and the collective space of the courtyard. Each of these entities as well as their implicit performative potential was thus clearly defined. Further, these entities were designed to create a gradient of devices of control: public control on the street, collective control over the courtyard, and individual control in the house.
From the early twentieth century on, the growing popularity of the garden city model introduced some ambiguity in the definition of the boundaries that prevailed hitherto in the historic city. Combining tropes of urban and rural life, the garden city model attempted to develop further the idea of community, a community of individuals, nonetheless, which bypassed the participation of the anonymous individual in the life of the city. Unsurprisingly, the figure of the street was clearly downplayed in this urban model. In

Dragon Court村庄_Dragon Court Village/Eureka
阿尔卡比德希的社会建筑群_Social Complex in Alcabideche/Guedes Cruz Arquitectos
Ulus Savoy住宅_Ulus Savoy Housing/Emre Arolat Architects
Version Rubis住宅_Version Rubis Housing/Jean-Paul Viguier et Associes

提高多孔性：重新思考住宅与社区之间的交界面_Enhancing Porosity: Rethinking the Interface Between the Dwelling and the Community/Nelson Mota

提高多孔性：重新思考住宅与社区之间的交界面

住宅与街道

两个世纪以来，一些影响城市设计发展的城市模型，表面上是试图包含且控制体量的。在欧洲，处在边界的建筑得到了广泛的认可，从19世纪晚期到20世纪上半叶，现代模型成为了处理工业区域的快速发展和解决人口激增所诱发的社会问题的最佳方案。这个模型的一个主要立面直接定义了公共同私人领域间的界限。而处在边界的建筑定义了街道、住宅和庭院公共空间之间的的急剧转变。所有实体中的任何一个（及其所表现的内在的潜在性）都进行了清晰的定义。此外，实体设计创造了层级式的控制装置：公众管理着街道，集体管理着庭院，个人管理着住宅。

从20世纪上半叶开始，日益盛行的花园城市模型模糊了风靡于历史名城中的边界的定义。结合了对城市和乡村生活的比喻，城市花园模型企图进一步发展社区理念，即一个个人社区，忽视了城市生活中不知名人士的参与。毋庸置疑，这个城市模型清晰地演示了街道特色。事实上，这个模型是城市向乡村拓展的重要的参考模型，恰如Peter G.Rowe指出的那样，绘制了一幅中部的乡村风景画。从20世纪20年代起，在街道与房屋间的传统关系中，理性主义者引进了厚板和塔，来作为替代选项。这些建筑类型成为城市肌理中自由、独立的部分，将提前几十年并最终渗透到居住区域的设计与规划当中。私人领域和公众领域间尖锐而又突然的过渡时期，成为了反街道运动的全盛期。

在20世纪的后半期，在追求更为人性化的居住设计方法之后，建筑师们企图向乡土传统学习。因此，厚板和塔支撑的功能主义城市模型，被决心引导个人更加亲近地面的模型所取代。换句话说，在之前的模型当中，房屋跟城市之间不自然的界限被柔和的界限所取代，在这样的界限内，个人和大众迫于交流，亲近地面。所谓的高密度的低层住宅模型因此展现在人们眼前，此类方式有可能对容纳民众的这一行为进行协调，而这种方式也协调了个人领域以及人们所属的社区空间，并最终赋予这些社区作为大型公众空间——城邦的一部分意义。

现如今，这些模型仍然是我们的城市肌理中的一部分，仍然是大众建造房屋最常用的模型，尤其是在特殊的施工当中，如城市复建、老人院，或者是特殊的项目。然而，人们可以发现回避规范特征的明显趋势，并且继续采用重新界定私人和公众之间限制的设计策略，继续热衷于重新考虑房屋和街道之间的界限。

共享边界

Dragon Court村庄是Eureka为日本的爱知县所设计的建筑。它是一个设计策略的典型案例，旨在追求多孔建筑。这个项目最引人注目的方面便是它一反传统地界定了街道和房屋之间的界限。实际上，设计师巧妙地将公众空间扩建到私人领域，创造了新的街道立面，在场地的内部展开。这项施工减少了场地占用中常见的分等级的设计定义。实际上，正如设计者所说的那样，这个项目迎合了那些在生活方式上鼓励探索共享边界的群体。这个项目利用模糊的图形，创造了开放的、半开放的以及封闭的多样性空间，以前所未有的方式相互连接，看起来像是对合理布局发出了挑战。这个规划的布局验证了设计师探索场地的地貌特性，以及利用它来打造高密度的低层住宅的能力。

Eureka的设计策略探索了有关协调房屋的个人特性以及社区的集体特性间的矛盾方法。在Dragon Court村庄中，每一所房子都是独一无二的。然而，这个村庄以九所房屋为一个单元形成了建筑综合体，并且其布局让人们很难分清住宅是从哪里开始的，到哪里结束的。它们的个性融入整体。项目的具体化将有助于进一步探索这个矛盾方法。一方面，木质的立面以其不同层次的透明度，使施工项目成为一个整体。另一方面，木质的框架界定了地面层带顶的开放空间的界限，暗示通过经费来实现个人表达的可能性。Eureka设计的Dragon Court村庄也因此成为住宅建筑群，这些建筑能够探索整体的孔隙度，从而缔结了个人与社区之间的条约。

住宅和纪念碑

在Dragon Court村庄中，个人单元融于整体，而在由Guedes Cruz建筑事务所设计的阿尔卡比德希的社会建筑群中，每个实体的个性是显而易见的。然而，为了突出群体的集体特性，每个单元都被连接起来。值得肯

fact, this model became a vital reference for the extension of cities towards the rural countryside, building a middle landscape, as Peter G. Rowe put it. From the 1920s on, the rationalists brought forth the slab and the tower as an alternative to the traditional relation between the house and the street. These building types, championed as freestanding and self-contained components of the urban fabric, would eventually pervade the planning and design of residential areas through many decades ahead. This was the heyday of the anti-street movement, a moment when the transition between the private realm and the public realm became sharp and abrupt.

Through the last half of the twentieth century there was an attempt to learn from the vernacular after a drive to pursue a more humanistic approach to the design of the habitat. As a consequence of this, the functionalist urban model supported by the figures of the tower and the slab was superseded by a model determined in bringing the individual closer to the ground, at it were. The so-called low-rise high density model thus surfaced as the possibility of accommodating the masses in such a way as to negotiate the realm of the individual with the community to which he or she belongs, and ultimately to make sense of these communities as part of a large public space, the polis.

Nowadays, these models are still part and parcel of the urban fabric of our cities and the most usual figures used in mass housing. In special operations such as urban renewal, elderly housing, or distinctive programs, one can nevertheless observe a noticeably tendency to shun away from canonical figures, and pursue design strategies engaged in redefining the limit between the private and the public, and keen in rethinking the boundaries between the house and the street.

Dragon Court村庄，爱知县，日本
Dragon Court Village, Aichi, Japan

定的是，出于隐私和安全原因，整个建筑群形成了一个封闭的社区。尽管排他性隐含在这种住宅区中，但建筑师旨在创造"隐私和社会生活之间的平衡"，即典型的地中海生活方式特色，体现在该地区的乡土传统中。实际上，作为地中海风格的乡村，在阿尔卡比德希老旧的社会建筑群里，人们可以看见房屋和扩建空间之间的智能连接，这种连接在整体的不同元素（街道、广场、小路和花园）之间创造了一种多孔的关系。进一步来讲，这个建筑群复制了传统地中海风格的村庄中可见的一些等级关系。实际上，中央建筑的形象是作为大教堂和市政厅的替代建筑而建立起来的，由此界定了纪念碑的规模，52所住宅单元围绕它而建，并且进行了延伸，形成了网状，创造了符合人体尺度的建筑，以及个人领域的框架。

项目的建筑用语和材料定义暗示着设计师们对创造抽象的作品的强烈兴趣，他们在调整额度为7.5m的基础上，进行严格构筑。然而，这种死板性却由每个单元及其周围环境中引入特色的地形来进行互补。覆盖每个住宅单元的多功能半透明屋顶进一步表达了私人生活和社会生活之间的这种细致的平衡。日落之时，点灯一盏，参与到为提高建筑群的安全而共同努力的行列之中。与此同时，屋顶也表达了它们的独特性，以致于在某种程度上可以运用灯光色彩的变换，作为个人预警系统。譬如，当任一单元有紧急的事情，灯光便从白色转换到红色。

强化多孔性

在伊斯坦布尔，Ulus Savoy住宅的布局同上述所探讨的阿尔卡比德希的社会建筑群的形态特质产生了共鸣，在这两个案例中，设计师们都企图利用地形来连接场地中不同的体量，从而营造意义非凡的间隙空间。说来也奇怪，在Ulus Savoy住宅项目中，在Emre Arolat建筑师事务所，Ertuğrul Morçöl，以及Selahattin Tüysüz接受委托来改造这个项目时，这方面便确定下来。他们不过致力于在房屋中间创造吸引人眼球的空间质量，并因此通过一个有意义的关系体系来探索建筑群的孔隙度。

设计团队接受了建筑法规和设计决策所规定的限制，这些限制事先

Sharing Margins

Dragon Court Village, designed by Eureka for the Japanese prefecture of Aichi, exemplifies a design strategy keenly engaged in pursuing a porous architecture. One of the project's most striking aspects is its unconventional definition of the boundary between the street and the houses. In effect, the designers shrewdly extended the public space into the private, creating a new street facade unfolding in the interior of the plot. This operation reduces the usual hierarchical definition of the plot's occupation. In effect, as the designers assert, this project caters for a community whose lifestyle praises the exploration of shared margins. The project plays with ambiguous figures and creates a multiplicity of open, semi-open and closed spaces that are articulated in unexpected ways, seemingly challenging a rational layout. The plan's layout testifies to the designers' ability to explore the morphological characteristics of the plot and take advantage of it to promote a low-rise high-density occupation.

Eureka's design strategy explores an ambivalent approach as regards the negotiation of the individual character of the house with the collective nature of the community. In the Dragon Court Village each house is unique. However, the nine housing units that form the complex are organized in such a way that one can hardy distinguish where each dwelling starts and where it ends. Their individuality is dissolved into the whole. The project's materialization further contributes to explore this ambivalence. On the one hand, the wooden facade, with its different levels of opacity, unifies the whole operation. On the other hand, the wooden frames that define the covered open spaces on the ground floor, suggest the possibility to accommodate individual expression through appropriation. Eureka's Dragon Court Village is thus a housing complex that masterfully explores the porosity of the ensemble to foster an engagement of the individual with the community.

The Dwelling and the Monument

While in the Dragon Court Village the individual unit is dissolved into the ensemble, in the Social Complex in Alcabideche by Guedes Cruz Arquitectos, the individuality of each entity is conspicuous. Each unit is nevertheless articulated in order to emphasize the collective character of the group. To be sure, for privacy and security reasons, the whole complex performs as a gated community. Notwithstanding the exclusivity implicit in this kind of residential complexes, the architects aimed at creating "a balance between privacy and life in society", a typical characteristic of the Mediterranean lifestyle, manifesting in the region's vernacular tradition. In effect, as in a Mediterranean village, in Alcabideche's elderly housing complex one can observe a smart articulation between the house and its extensions that creates a porous relation between the different collective components of the whole ensemble: the streets, squares, paths, and gardens. Further, this complex replicates some of the hierarchical relations observed in the traditional Mediterranean village. In effect, while the figure of the central building stands as a surrogate for the Church or the Town Hall, thus defining the monumental scale, the fifty-two dwelling units which spread around it entwine the net that creates the human scale and frames the realm of the individual. The architectural language and material definition of the project suggests a keen interest in creating an abstract composition, rigidly defined by a strict composition based on a modulation of 7.5 meters. This rigidity is nevertheless compensated with situated responses to the topography that introduce some distinctiveness in each unit, and in the spaces surrounding it. The multifunctional translucent roofs that cover each dwelling unit further express this delicate balance between private life and the life in society. They light up at the end of the day, thus participating in a collective effort to enhance the security of the whole complex. At the same time, they express their singularity to such an extent

阿尔卡比德希的社会建筑群，卡斯凯斯，葡萄牙
Social Complex in Alcabideche, Cascais, Portugal

获得批准，以引入新的材料，符合对整体建筑群来说较高的质量需求。新设计探索了住宅同其间的上空体量之间的明显对立性。建成体块的毫无特色的形态特征与开放空间的动态特色形成互动，这种互动提高了建筑师新建的地形（探索集体空间，将其作为城市社区中的一个重要元素）的重要性。

重新定义边界

上述案例中所讨论的孔隙度带来了空间品质，首要由各个单元（一个单独住所或住房体块）之间的相互作用来形成，同时也与住宅整体相关。但在接下来要介绍的项目中，情况都是不同的。

在Jean-Paul Viguier et Associes为蒙彼利埃市所设计的Version Rubis住宅项目中，其最显著的特点是在私人空间领域及其直接扩建领域之间建造过渡空间的推动力，这种推动力渗透到项目的每一个角落中。这个特色在建筑的立面设计中得到了完美的表达。设计师们建造了一个厚膜，包围了整个建筑，并且通过它的四个面来不断变化。

在西立面的悬臂式阳台中，外围护结构极尽表现主义的时尚之美。阳台的扭曲投影朝向阳台之外，试图故意将住宅的私人领域与街道的公共性质形成对比。这些元素所投射的张力，试图挑战和重新定义内部和公众空间之间的典型界限。

协调住所和城市

在接下来的项目中，我们可以看到，这些建筑项目故意试图挑战风靡于20世纪的权威的房屋模型。这些项目以不同的方式，来调整矛盾和歧义，致力于重新积极考虑私人住所的特性和城邦的公共特性之间的关系。因此，这些项目实例努力提高案例中个人住宅和集体建筑的孔隙率，同那些试图重新界定房屋和街道的界限的例子一样，我们能够发现这样的住宅建筑的实例，即个人对社区生活的政治本质发出挑战的建筑实例。

that they can be used as an individual alert system – changing the colour of the light from white to red – whenever an emergency happens in one of the units.

Enhancing Porosity

The layout of the Ulus Savoy Housing, in Istanbul, resonates with the morphological qualities of the Social Complex in Alcabideche, discussed above. In both cases there is an attempt to take advantage of the topography to articulate the different volumes on the plot and to create meaningful interstitial spaces. Curiously enough, in the Ulus Savoy Housing this aspect was already defined when the team of Emre Arolat Architects, Ertuğrul Morçöl, and Selahattin Tüysüz were commissioned to revise the project. They nevertheless contributed to introduce a noticeable spatial quality to the spaces in between the housing blocks and they thus explored the porosity of the complex through a meaningful system of relations. The design team accepted the constraints triggered by building regulations and design decisions approved in beforehand to introduce new material and expressive qualities to the whole complex. The new design explores an explicit confrontation between the housing blocks and the voids in between them. The interplay between the banal morphological characteristics of the project's built mass and the dynamic nature of the open space enhances the importance of the new topography "invented" by the authors to explore the collective space as a vital component of an urban community.

Redefining Boundaries

The spatial qualities brought about by the porosity discussed in the cases examined above were, first and foremost, triggered by the interplay between the individual unit (a single dwelling or a housing block), and the whole housing ensemble. In the following project the situation is different.

In the case of the Version Rubis Housing, designed by Jean-Paul Viguier et Associes for Montpellier, the project's most notable feature is its pervasive drive to create spaces of mediation between the private realm of the house, and its immediate extensions. This is utterly expressed in the design of the building's facades. The designers created a thick membrane that envelopes the building and changes constantly through its four sides. This envelope reaches an expressionistic fashion in the cantilevered balconies located at the west facade. The distorted projection of the balconies to the outside suggests a deliberate attempt to confront the private realm of the house with the public nature of the street. The tension projected by these elements, expresses an attempt to redefine and challenge the typical boundaries between the domestic and the public space.

Reconciling the *domus* and the *polis*

In the projects featured in the following pages one can observe a deliberate attempt to overcome the canonical housing models that pervaded through the twentieth century. In very many different ways, these projects accommodate ambivalence and ambiguity as positive contributions to rethink the relation between the private character of the domus and the public nature of the polis. Hence, both in the case of the projects that strived to enhance the porosity between the individual and the collective as in those cases that attempted to redefine the boundaries between the house and the street, we can find instances of an architecture of dwelling that confronts the individual with the political nature of life in community. Nelson Mota

Dragon Court村庄

Eureka

低密度住宅,面向周边地区和环境开放

Dragon Court村庄是包含九个住宅单元的出租排屋。场地位于可供居住的郊区内,这里的每个居民都拥有一辆汽车,实现了机动化。每个家庭都拥有一个停放两辆车的停车场,且场地的一半面积都被停车区和车道占用。因此,房屋总面积远远小于法规规定的面积。建筑师旨在建造一处暴露在周边地区和环境的住宅区,且每个单元之间都有公用的边界。

通透环境周围的小巷式空间

这是一座通透的建筑,微风和树荫赋予它一个舒适的空间系统。这座建筑在周围设有边界,由一条环形小巷和停车区界定。体块利用允许半户外环境贯穿整座建筑的风力模拟来进行调整。由于建有半户外空间和附属建筑,建筑内的生活可以扩展到户外空间,甚至扩展到小巷和整个区域内。在未来,由于老龄化社会和汽车共享理念的渗透,汽车的数量便会减少,而停车场也将被用作花园,或者是遛狗场地。建筑将会被面向社区开放的活动所簇拥。

不同的分区和共享的屋顶户外空间

整个体块是由五个缠绕在一起管状结构构成,使人们看不见整体建筑的场景。这个组装结构拥有九处空间,各有相似和差异之处,共同组成一座大型独立住宅。此外,附属建筑设在单元之间,木框架固定在整个体块中,平台设在一层,以再次模糊每个住宅单元之间的界限,通过这些重叠的区域,这个项目的规划形成了不同规模的共享空间。

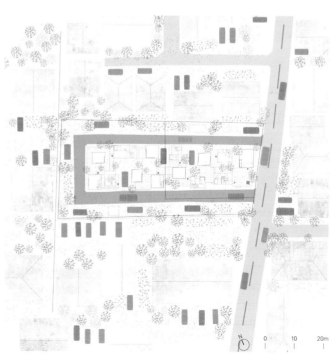

住宅组群的可持续建造方法

在亚洲的城市和村庄进行场地工作,面对恶劣的天气和全球变暖问题所产生的日益严重的威胁,建筑师有时要面对能够适应自然灾害以及常用的建筑行为的住宅文化,而这些行为都与环境和方法产生联系,以助于维持这一状况。

同时,观察本地传统的建筑和村庄的行为能够与观察这些建筑和村庄的拆毁和接受度取得同样的效果。

这里几乎从未有过的贫困生活,有时要受到带有不符合规定的密度的周围环境的制约,同时由于渐次上升的空间品质(换句话说,建筑过剩且不断更新的自然属性),项目设计会更依赖生活的不确定性。在这一方案中,这些品质都得以采纳。

例如,这些品质可以在主要房间和屋檐下的空间的下方看到。它们不时地被共享和拥有。此外,建筑师相信空间管理,如木框架对环境进行调整,或者将边界看作是潜在的可拓展空间,都能实现一个可持续的住宅建筑群。

Dragon Court Village

Low density residence that opens to the surrounding area and environment

Dragon Court Village is a rental row house comprised of nine units. The site lies in a motorized society of a residential suburb, where each person owns a car. Each household has a parking area for two cars and half of the site is occupied with parking areas and driveways. Thus, the total floor area is much less than the one that the regulation allows. The architects aimed for a residence exposed to the surrounding area and environment with a shared margin between each unit.

An alley-like space surrounding a porous environment

This is a porous building, with the breeze and shade leading it to a comfortable spatial system. The building has a margin around

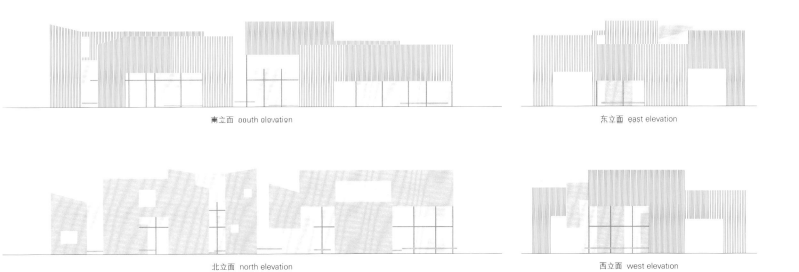

南立面 south elevation 东立面 east elevation

北立面 north elevation 西立面 west elevation

项目名称：Dragon Court Village
地点：Aichi, Japan
建筑师：Eureka
主要负责人：Junya Inagaki, Satoshi Sano, Takuo Nagai, Eisuke Hori
项目团队：Kazutoshi Sugimoto, Yuki Nagasawa, Hiroyuki Tsukada, Kazunori Yamaguchi
顾问：Takuo Nagai　结构：Eisuke Hori
环境管理：Kiyofumi Kobayashi, Satoru Mori
总承包商：Taikei Construction Co., ltd.
用地面积：1,177m² / 总建筑面积：360m² / 有效楼层面积：508m²
结构体系：timber
材料：laminated veneer lumber, red lauan, mortar, galvanized steel, plaster board, oak flooring
设计时间：2012—2013 / 竣工时间：2013
摄影师：©Ookura Hideki(courtesy of the architect)

A-A' 剖面图 section A-A'

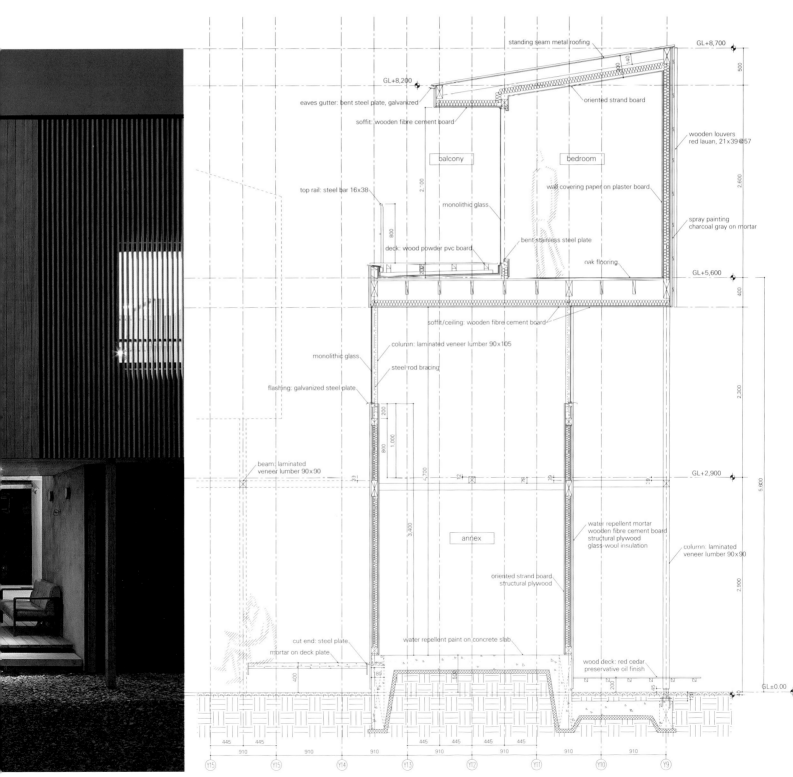

a-a' 剖面图 section a-a'

the rim, created by a circular alley and parking areas. The masses were adjusted by using wind simulations allowing semi-outdoor spaces to penetrate the building. Due to the semi-outdoor spaces and the annexes, life in this building will expand to the outdoor spaces and even to the alleys and the whole region. In the future when the number of cars will decrease as a result of aging society and the diffusion of car sharing, the parking areas will be used as things such as a garden or a dog-run place. The building would be surrounded by such kind of activities that opened to the community.

The diverse zoning and sharing of roofed outdoor space

The entire mass is structured with five tubes tangled together, denying an overview of the building. As well as being an assembly of nine spaces with both similarities and differences, they comprise one single massive house. In addition, the annexes placed between the units, the wooden frames fixed across the masses and the platforms settled on the ground blur again the boundaries of each unit of the residents. By these overlapping zones, it is planned to induce a diverse scale of sharing.

A sustainable method of group housing

Within the fieldwork in Asian cities and villages, conducted against the backdrop of increased threats from severe weather and global warming, the architects sometimes encounter housing cultures that adapt to natural disasters, customary architectural behaviors that have affiliate with the environment and methods to help maintain such things.

At the same time, the act of observing native and traditional architectures and villages is synonymous with observing their very own destruction and acceptance.

Life that barely scrapes by is sometimes subject to unlawfully dense surroundings and is very much reliant of the ambiguity of life due to the gradational spatial qualities. In this scheme, such qualities were adopted.

They can be seen for example, beneath the main rooms and under the space under the eaves. From time to time, they allow themselves to be shared or occupied. Furthermore, the architects believe that the spatial management such as environmental adjustments done by the wooden frames, or the potential for extensions that the margins behold, can realize a sustainable residential complex.

室内透视图
interior perspective

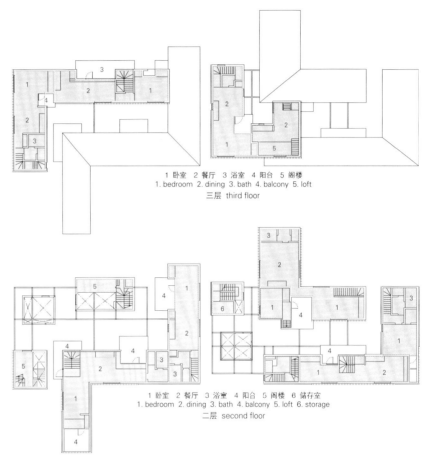

1 卧室　2 餐厅　3 浴室　4 阳台　5 阁楼
1. bedroom 2. dining 3. bath 4. balcony 5. loft
三层　third floor

1 卧室　2 餐厅　3 浴室　4 阳台　5 阁楼　6 储存室
1. bedroom 2. dining 3. bath 4. balcony 5. loft 6. storage
二层　second floor

1 入口　2 附属建筑　3 餐厅　4 浴室　5 卫生间　6 储存室　7 停车场
1. entry 2. annex 3. dining 4. bath 5. W.C. 6. storage 7. parking
一层　first floor

住宅单元类型
house unit type

单元1 unit 1
单元2 unit 2
单元3 unit 3
单元4 unit 4
单元5 unit 5
单元6 unit 6
单元7 unit 7
单元8 unit 8
单元9 unit 9

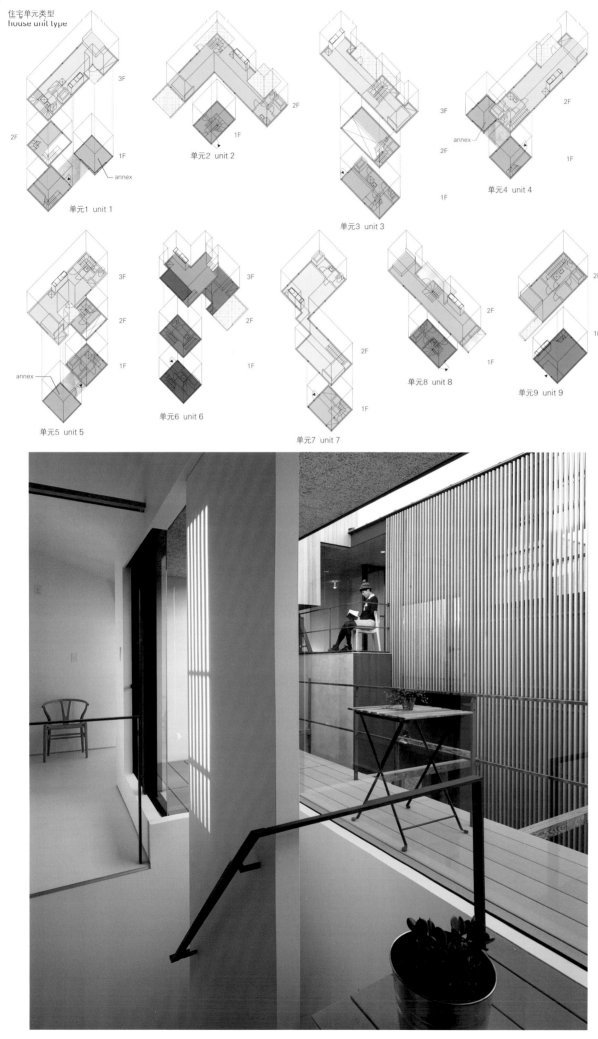

阿尔卡比德希的社会建筑群

Guedes Cruz Arquitectos

阿尔卡比德希的社会建筑群位于里斯本的卡斯凯斯市区，总建筑面积约为10 000m²，旨在重建一种地中海式的生活方式，在这里，街道、广场以及花园的室外空间就像房子本身的扩建结构一样。

项目于2012年竣工，包含52所住宅（为老年夫妇建造的住宅单元），以及一个主要的服务大厦，容纳了社交区域、带有私人房间的护理区域，以及卧床不起的病人区。如同在麦地那城市一样，相互连接的布局的调整额度为7.5m，拥有为行人而设的不同宽度的街道，而行人们在白天也享受住宅所提供的遮阴保护，且晚上被住宅内发出的光线所引领。

在这个项目中，光线是一种享受。半透明的屋顶发出的光线在一天结束的时候照亮，十个屋顶为一组，覆盖在建筑群之上，巧妙且均匀地照亮了街道、广场和花园。一种平静但轻快的氛围允许用户夜晚在建筑群内的不同层次和空间走动，无需担心，也没有任何限制。而发生紧急事件时，用户也可以启动报警系统，向位于中央建筑的控制站报告，且箱形的屋顶所发出的光线也有所变化：大型白色箱形结构变为红色。整个建筑群的可持续性和环境的平衡起源于对最初的投资和运营成本之间的关系的体系性思考，并且意识到后期资金的减少意味着前期投入的增加。

最终的解决方案取决于其形式的有效性或者系统的功效和要选用的材料。清水混凝土和有机玻璃是该建筑群的主要材料。因为房屋是这个建筑群的主要结构元素，所以其理念也是设计和操作原则的基础。房屋由清水混凝土构成，带有一个纹理粗糙的木质板条框架，容纳了居住区、起居室和厨房、卧室、浴室，以及一个箱形结构，这个结构的材料，即有机玻璃，具有特殊的性质：能够反射（白色的）太阳光线，具有防透性，能够抵抗恶劣的天气，防火，但允许光线进入（玻璃为半透明材质）。

住宅的室内温度比较稳定，这是屋顶的箱形结构的反射性能与有机玻璃箱和混凝土基座之间形成的气垫所发挥的热效应所致。中央建筑也采用同样的造型和原则，包含了所有的公共服务区，以发挥必要的功能，提高生活质量。它设有电气生产系统、太阳能光伏发电厂，二者均能为可居住的住宅区提高援助。

Social Complex in Alcabideche

Located in Cascais, in the metropolitan area of Lisbon, the Social Complex of Alcabideche with a total construction area of approximately 10,000m² aims to reconstitute a Mediterranean life style in which the outdoor spaces of streets, plazas and gardens are like an extension of the house itself.

Concluded in 2012 it comprises 52 houses (housing units intended for elderly couples) and a main support building that houses the social areas, a nursing area with individual rooms and an area for bedridden patients. The connective layout with a modulation of 7.5m has, as in Medina, streets of different widths reserved for pedestrians who enjoy the protection of the shade provided by the houses by day and at night are guided by the light the houses give off.

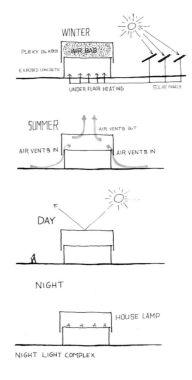

服务大厦
service building

三层 third floor

二层 second floor

1 独立房间	1. single room		
2 护理区/辅助浴室	2. nursing/assisted bath		
3 食堂	3. refectory		
4 厨房	4. kitchen		
5 男士卫生间	5. male W.C.		
6 女士卫生间	6. female W.C.		
7 理发店	7. hairdresser		
8 接待处	8. reception		
9 女士盥洗室	9. female restroom		
10 男士盥洗室	10. male restroom		
11 体育馆	11. gym		
12 泳池	12. pool		
13 带顶的平台	13. covered deck		
14 员工室	14. staff room		
15 男性员工更衣室	15. male staff dressing room		
16 女性员工更衣室	16. female staff dressing room		
17 储存室	17. storage		
18 音像图书馆/机房	18. video library/Internet		
19 电梯和楼梯	19. lifts and stairs		
20 交通流线	20. circulations		
21 技术区	21. technical area		
22 医用电梯	22. bed lift		
23 停车场	23. parking		
24 双人房	24. double room		
25 主接待区	25. main reception		
26 大众入口	26. general entry		
27 医疗护理区	27. medical care		
28 起居室/酒吧	28. living room/bar		
29 室外平台	29. outdoor deck		
30 卧床区	30. bedridden room		
31 员工室	31. staff room		
32 会议室	32. meeting room		
33 行政办公室	33. administration office		
34 行政接待区	34. administration reception		
35 董事会会议室	35. boardroom		

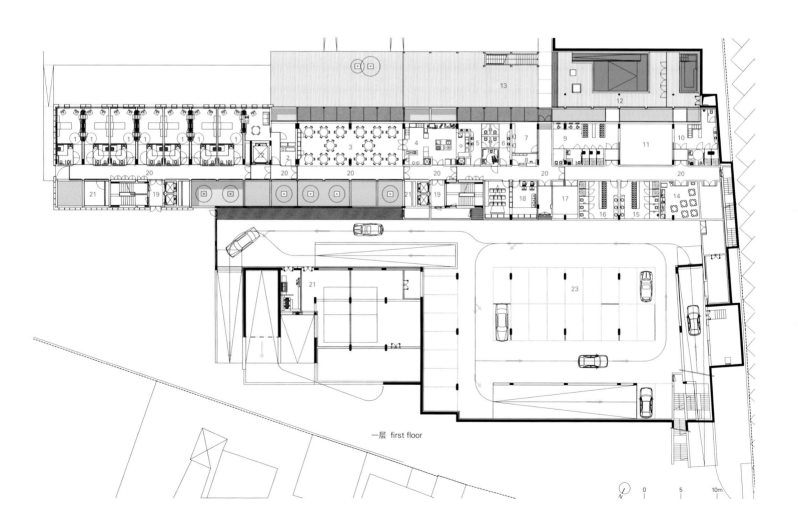

一层 first floor

In this project light is joy. The translucid roofs light up at the end of the day in groups of 10, alternately, over the area of the complex, subtly and evenly lighting up streets, plazas and gardens. A calm but cheerful atmosphere is created that allows users to circulate at night among the different levels and spaces of the complex without worries or constraints. In the event of an emergency, users can activate an alarm that alerts the control station located in the central building and the box-shaped roof's light changes: the large white box turns red. The sustainability and environmental balance of the whole complex stems from the systematic consideration of the relationship between the initial cost of the investment and its operating costs, being aware that a reduction in the later normally requires an increase in the former.

The final solutions depend on the effectiveness of its form or on the efficacy of the functioning of the choices of systems and materials to be used. Exposed concrete and plexiglass are the main materials in the Complex's architecture. As the houses are the structural elements of the whole complex it was on their conception that the architects based the design and operational principles. The houses are formed by a nucleus of exposed concrete with a "rough" formwork of wooden lathes that house the habitable area, living room and kitchen, bedroom and bathroom and a box in which the material, plexiglass, has specific characteristics: reflection of the sun's rays (white), being impermeable, being resistant to the weather and to fire and allowing the light in (translucid).

The stabilization of the temperature on the inside of the housing units is due to the reflective properties of the white box of the roof and the thermal efficiency of the air cushion created between the plexiglass box and the concrete base. The central building, within the same modelling and principles, contains all of the common services necessary for proper functioning and quality of living. It includes the electricity production systems and the photovoltaic plant, which supply the habitable areas of the residence.

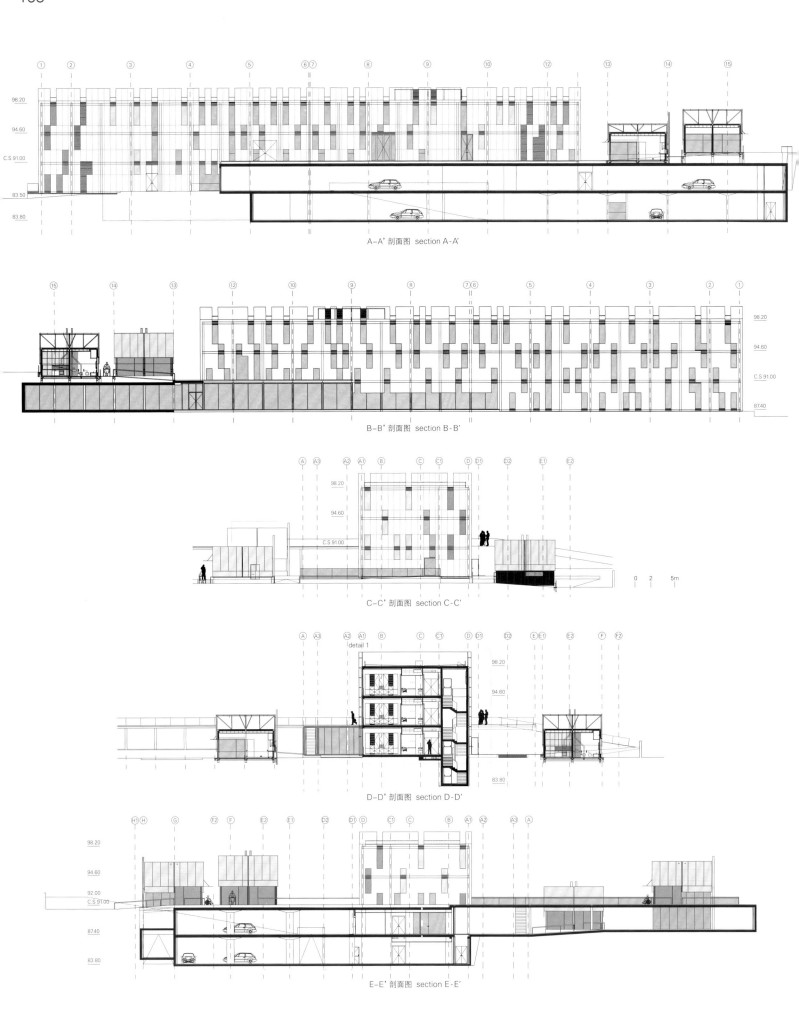

A-A' 剖面图 section A-A'

B-B' 剖面图 section B-B'

C-C' 剖面图 section C-C'

D-D' 剖面图 section D-D'

E-E' 剖面图 section E-E'

项目名称：Social Complex in Alcabideche
地点：Alcabideche, Cascais, Portugal
建筑师：Guedes Cruz Arquitectos
项目团队：José Guedes Cruz, César Marques, Marco Martínez Marinho
发起人：Fundação Social do Quadro Bancário
合作者：Patrícia Maria Matos, Nelson Aranha, João Simões, Isabel Granes
结构工程师：PPE
特殊设备设计师：Espaço Energia
景观建筑师：Paula Botas
施工联盟：FDO + JOFEBAR
监理：Mace
用地面积：12,876m²
有效楼层面积：9,956m² (2,756m²_housing unit, 7,200m²_main building)
单元面积：53m²
竣工时间：2012
摄影师：©Ricardo Oliveira Alves (courtesy of the architect)

详图1 detail 1

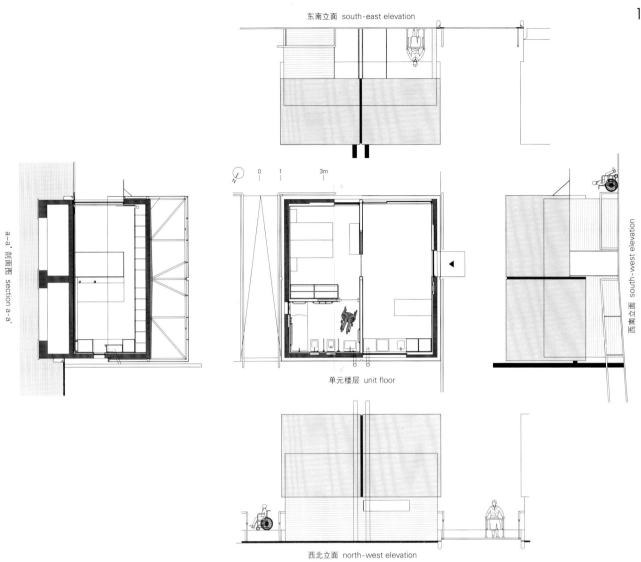

Ulus Savoy住宅

Emre Arolat Architects

该项目位于80 000m²左右的Ulus场地内，包括26个相同大小的地块，因此项目需要两项重要的投入。第一个投入是建筑环境。第二个投入，它作为实体项目出现在我们面前，是事先准备好的，其错综复杂的审批过程也都是经合法程序完成。为此，在这种背景下，投资者更倾向于保护这些条款。新设计完全保留了现有项目的数量、位置和建筑体块的水平面，并将其作为一种要求。而决定了设计进程的最重要的因素便是这一要求，结合建设投入（定义相当严格的，即将建造的大型体块结构），如15m×20m的基础区域，与场地的斜坡、一层以上的楼层规划、四边带有33°斜坡的屋顶相平行。这个项目所面临的另一个紧急的情况是谈及Ulus山谷显著的建筑特性是毫无意义可言的。大量封闭的社区似乎已经渗透到每一个能见的角落和缝隙，一个依托着另一个，所有的社区都是在上述的同等条件下建成。

车库问题优先成为设计的主要条件元素之一。车库的水平位置被设计得能够容纳足够多的汽车，并且适合每座建筑的地下室。坡道连接了车库的不同层次，因此使其变得很流畅。一些覆盖屋顶的外壳完全移除；此外，两个外壳之间的不同水平面产生的缝隙，通过模糊地下层和外部的边界，来形成微小的差别，十分奇妙。这些差别可使日光穿透外壳结构，将休闲区域和外部区域相联系，同时也旨在晚上由于光线会从内部渗透，也可以用来照亮景观。

设想一下，外壳将是定居于此的一个重要特性，主要应用在室外景观中，会在某些地方呈现出它们的硬度和锐度，尽管大部分经常被"隐藏"于植被层中，并且被生长的植物漫过，但是它们仍然可以在该区域的所有地方合理地存在着。

项目旨在展示结构，一个支离破碎的统一体，没有失去它的连续性，沿着车库、外部景观和住宅体块贯穿于整个表面，同时，其模式也将形成并且转化为连接的纽带，以防止由于全方位的映射效应而在地形和结构之间产生分离。

Ulus Savoy Housing

There were two important inputs regarding the settlement located in Ulus, which was to be situated on a lot of approximately 80,000 square meter and would consist of 26 masses of the same size. The first of these inputs is the building conditions. As for the second input, it emerged before us as the entity of a project that was prepared beforehand, whose intricate processes of approval were legally completed, and for which, within this context, the investor preferred to preserve the outlines. The number, location and levels of building blocks in the existing project were exactly preserved in the new design emerged as a requirement. The most important factor that determined the course of the design was this requirement, combined with building inputs, which defined quite rigidly the massive structure of the blocks to be built, such as the 15-by-20-meter base area of blocks that was to be parallel to the slope of the lot, the projections on stories above the

ground floor, and the roofs with a 33-percent slope on all four sides. Another of the project's exigent features was the fact that it is meaningless to speak of the noteworthy architectural characteristic as regards the Ulus Valley, which has become a torrent of buildings due to the many gated communities that seem to have penetrated into every nook and cranny that could be found, side by side, one on top of the other, all built under the same conditions mentioned above.

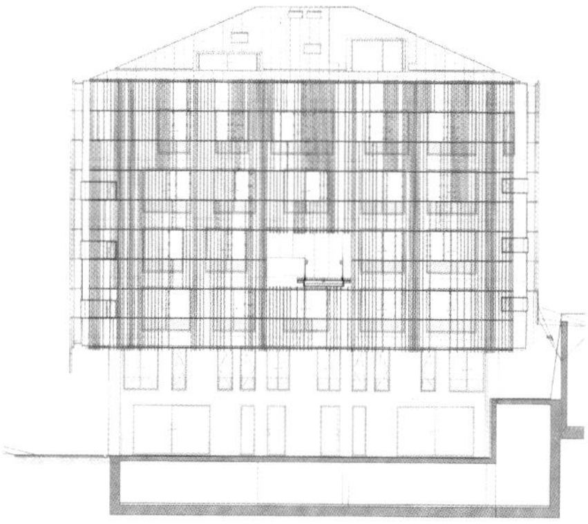

A-A' 剖面图 section A-A'

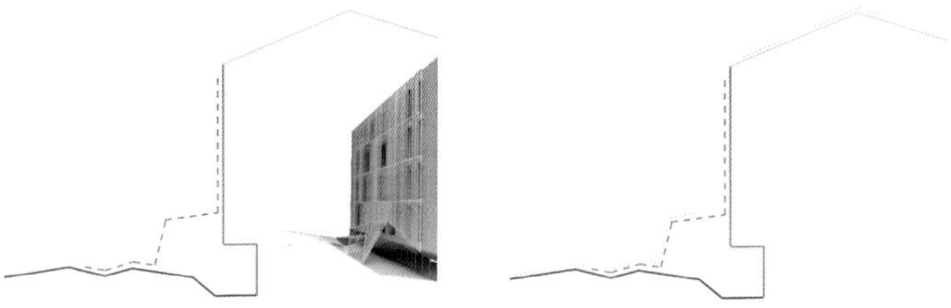

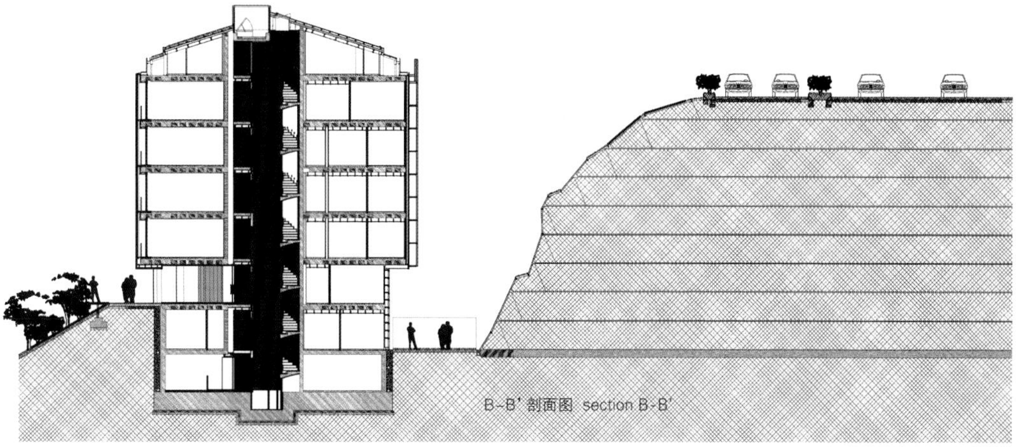

B-B' 剖面图 section B-B'

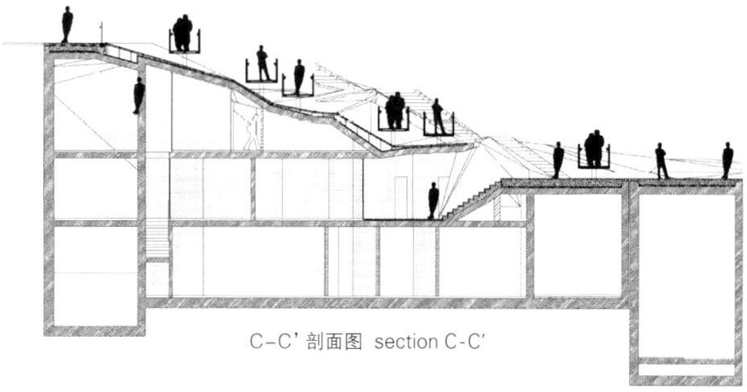

C-C' 剖面图 section C-C'

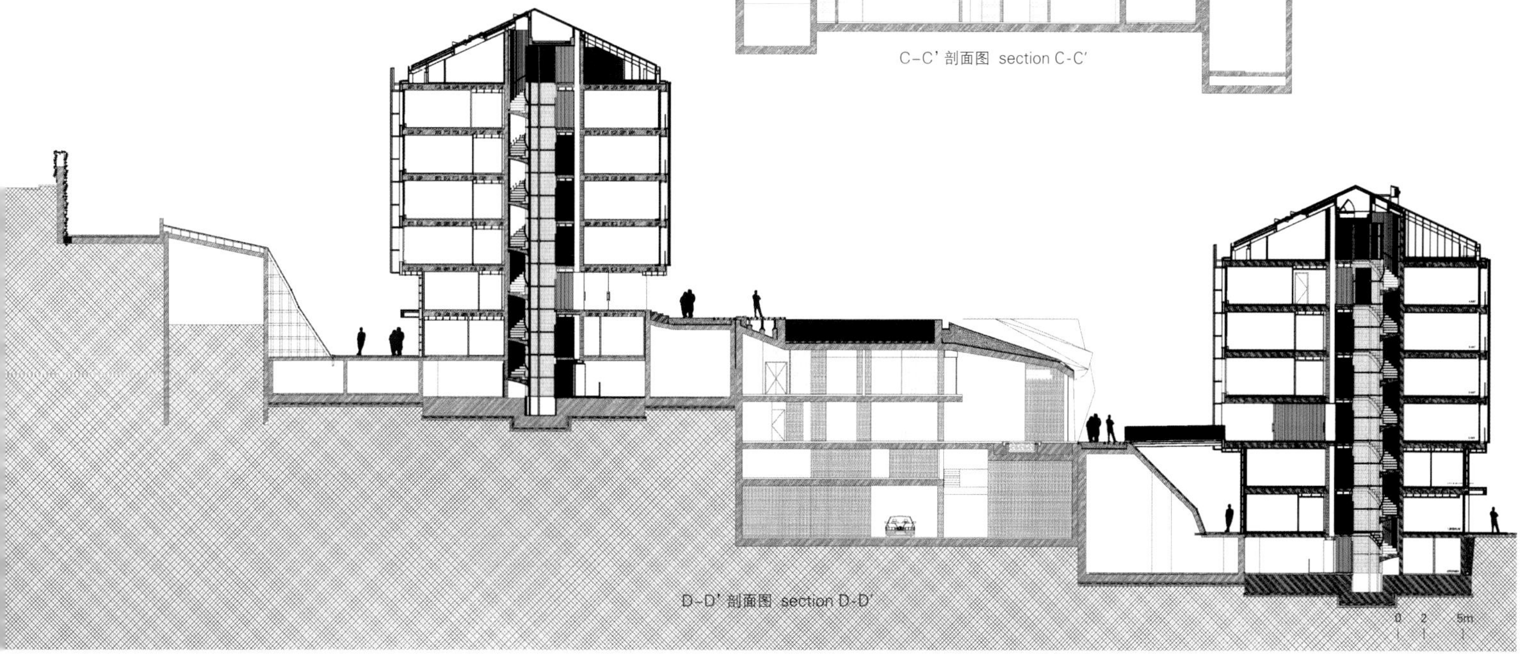

D-D' 剖面图 section D-D'

The garage issue was given precedence as one of the main conditioning elements of the design. The garage level was designed to hold a sufficient number of cars and to fit the basement of each building block, by problematizing the upper cove. Each different level in the garage was connected by ramps, thus rendering it fluid. Some of the shells that formed the cover were completely removed; added to this, occasional slits and interstices formed by the slight difference in level effected between two shells created surprising nuances by blurring the boundary between the underground layer and the exterior. These nuances enabled both daylight to penetrate the shell structure, and the connection between recreational areas and the exterior. The aim was also that at night they be used to illuminate the landscape thanks to the light that would seep from the inside.

It was envisaged that it would be an important characteristic for the settlement that the shells, which are the main material of the exterior landscape, would, in places, be present in all there hardness and sharpness, though most are often partly "concealed" by the layer of vegetation growing over them, while they would still remain legible in some way from all points of the area.

The aim was that the fabric is present, though with a fragmentary identity and without loosing its continuity, throughout the entirety of the surfaces running along the garage, exterior landscape, and housing blocks; and also that the pattern which would take shape therefrom would turn into a coupling agent that would prevent disjunctions between the topography and the structure thanks to its all-extensive mapping effect.

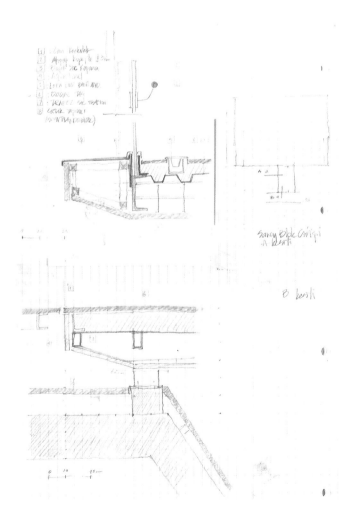

项目名称：Ulus Savoy Residences
地点：Istanbul, Turkey
建筑师：EAA-Emre Arolat Architects
理念设计：Emre Arolat, Ertuğrul Morçöl, Selahattin Tüysüz
结构方案：Gonca Paşolar, Kerem Piker, Deniz Kösemen, Zeki Samer, Serdar Sipahioğlu
责任建筑师：Gonca Paşolar
管理员：Deniz Kösemen
项目团队：Gülseren Gerede Tecim, Zeynep Yapar, Nesime Önel, Sezer Bahtiyar, Olcay Özten, Volkan Yağ, Özge Çağlayan, Hale Ikizler, Merve Yüksel, Süleyman Yıldız, Başak Tekin, Nida Pelin Üye, Sevim Uyan, Elif Ekim
建造商：Bahadır İnşaat
结构工程师：Ismet Babus Mühendislik
电气工程师：Detay Mühendislik
机械工程师：ÇağYapı Mühendislik
景观建筑师：Ds Mimarlık
甲方：Çarmıklı Saruhan Partnership
地块面积：59,900m²
有效楼层面积：83,557m²
停车场面积：17,752m²
设计时间：2005—2009
施工时间：2011—2013
摄影师：©Cemal Emden(courtesy of the architect)

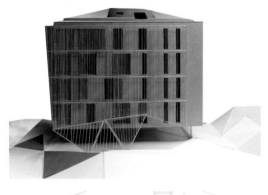

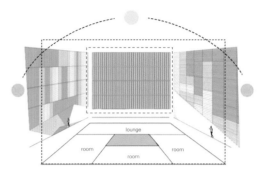

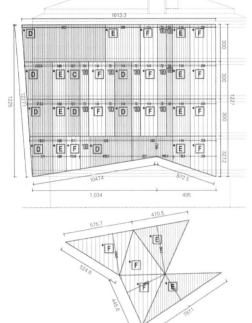

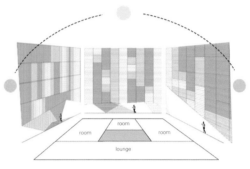

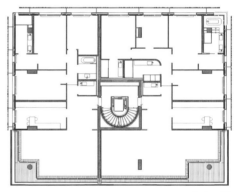

典型楼层平面 typical floor

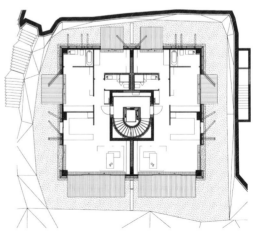

典型一层平面 typical ground floor

Version Rubis住宅

Jean-Paul Viguier et Associes

这座建筑与ZAC（混合型开发区域）的整体设计主题相契合，采用了一个整体的可持续发展方法：使用外部光线产生的阴影，来提供有效的保护，免受季节中期和夏季的太阳能所带来的伤害，同时却没有阻挡日光的射入。

为了与ZAC的城市和建筑规则一致，建筑包含一个模压的混凝土平台，三层至七层的主要结构位于其上，建筑同时还设有一个嵌壁式阁楼，整座建筑向上抬起，北面Mondial98号大街一侧为十层，而南侧为六层，不同的高度是为了阶梯式的顶层而设计的。

一栋五层的主楼是ZAC规则中四种推荐建筑的首选，以减轻"箱形"体量，这个"箱形"体量由顶部阁楼层构成（两层包含两个大型豪华公寓）。

这座建筑的体量十分复杂，由一系列凹进处和悬挑结构构成。所有的住宅都设有一个阳台或者一个露台，它们被固定的或者是可以推拉的嵌板覆盖；整座建筑是一个连续的体量，带有红色饰面（金属覆层，铝质盒形嵌板）。

该项目不但满足了ZAC的建筑规则，同时也满足了坚固、清晰的住宅体块特征，致力于向奢华靠拢，同时呈现出不同的建筑类型。

Version Rubis Housing

Following the overall theme of the ZAC (mixed development zone), the building adopts an overarching sustainable development approach: the use of external sun shades provides effective protection from mid-season and summer solar inputs, but without blocking natural daylight.

In keeping with the urban and architectural rules of the ZAC, the building comprises a stamped concrete podium upon the main portion of the building from the 3rd storey to the 7th storey sits, and a recessed attic. The whole building rises, at its highest point, to 10 storeys on the north side on Avenue du Mondial 98, and to 6 storeys on the south side, with varying heights for the "terraced" top storeys.

A five-storey main building was preferred to the four recommended by the ZAC rules, in order to lighten the volume of the "box"

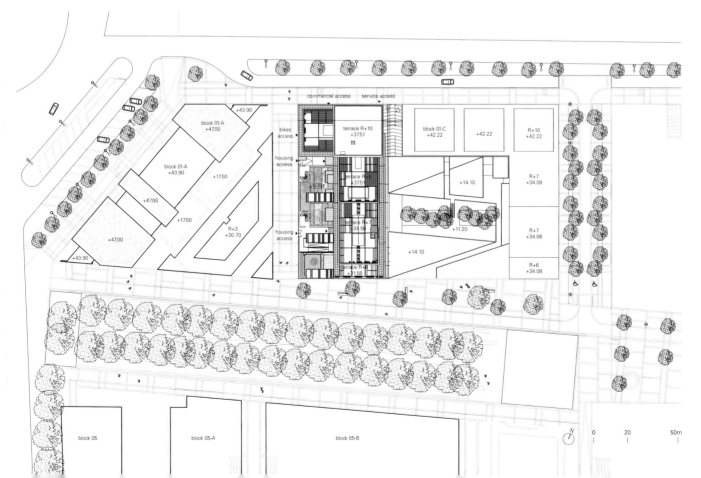

西立面 west elevation

东立面 east elevation

1. architectural concrete (as recommended by the ZAC)
2. white coating 3. fiberglass panels (random color variations: red 4 types)
4. sliding shutters with perforations fiberglass (random variations rewinder: 4 types of red)
5. cowling fiberglass or metal 6. railing galvanized steel gray 7. galvanized steel structure
8. sunscreen: store horizonal silver gray exterior 9. sunscreen: dark gray exterior vertical blind
10. solar panels 11. siding tolle or gray canvas

formed by the top attic floors (two floors containing two large luxury apartments).

The building has a complex volume formed by a series of recesses and overhangs. All homes will have a balcony or a terrace, protected by fixed or sliding panels: the whole constitutes a consistent volume finished in a red color palette (metal cladding, in aluminum cassette panels).

While satisfying the rules of the ZAC, this project is a block of dwellings with a strong, clear identity which contributes a touch of luxury amid the diverse architectural styles.

项目名称：Housing "Version Rubis", ZAC Parc Marianne
地点：Montpellier, France
建筑师：Jean-Paul Viguier et Associes
结构工程师：André Verdier / 流体工程师：ENR Concept / 认证：BBC Effinergie
表面积：6,000m² / 房屋总面积：11,000m² / 造价：EUR 7.5 M + tax
竞标时间：2009 / 获奖情况：Pyramides d'or 2012 / 设计时间：2009 / 施工时间：2011—2013
摄影师：©Takuji Shimmura (courtesy of the architect)

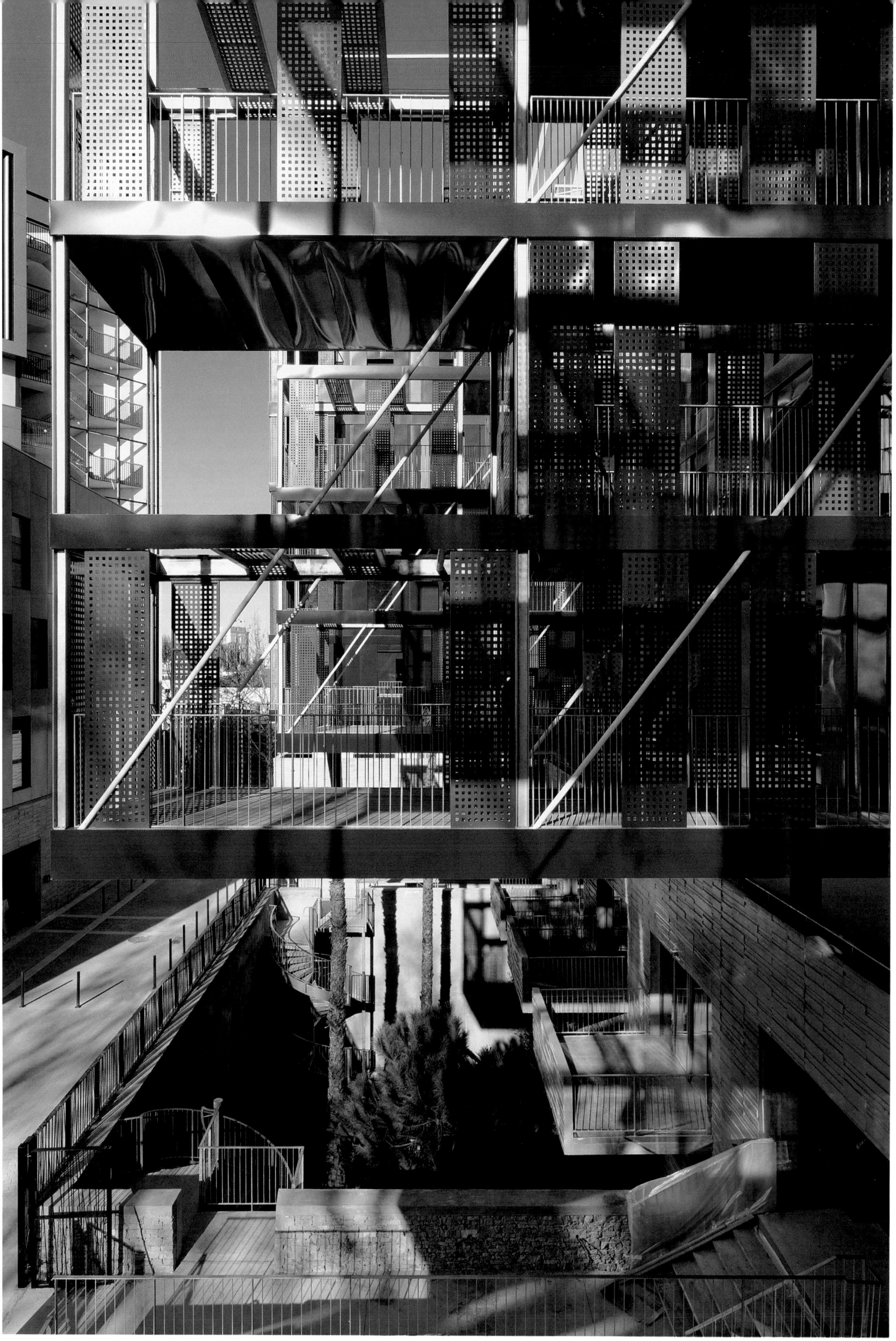

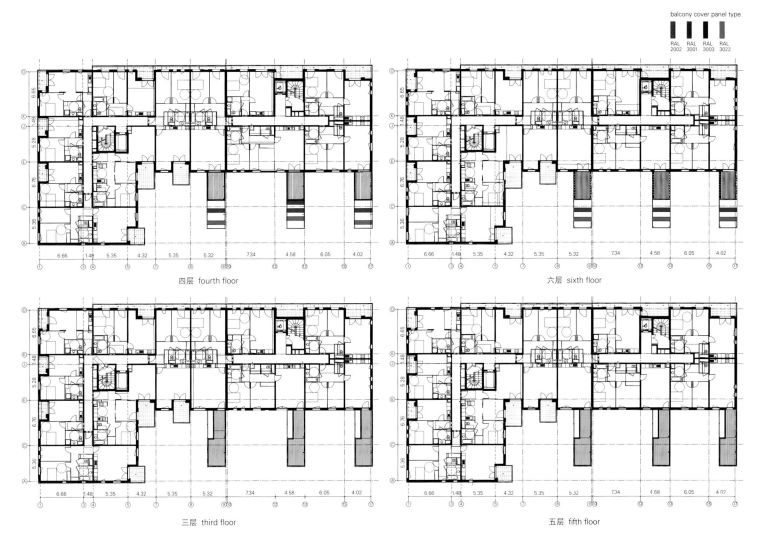

四层 fourth floor

六层 sixth floor

三层 third floor

五层 fifth floor

>>106
OAB-Office of Architecture in Barcelona
One of the founders Carlos Ferrater was born in 1944, Barcelona. He received a degree in Architecture in 1971, and his doctorate in 1987. He is a professor of architectural project design at the Polytechnic University of Catalonia. His work is known as effective combination of nature and geometry that does not sacrifice functionality. Núria Ayala joined Carlos Ferrater Studio 2000 and graduated from ETSAB in 2001. Even after the establishment of OAB in 2006, she has been collaborating with the firm as project director, collaborator, and co-author participating many kinds of different works, projects and competitions.

>>16
spacialAR-TE
Álvaro Fernandes Andrade received a B.Arch in 1999 from FAUP(Oporto School of Architecture) and a Master in Planning and Project of Urban Environment from FAUP and FEUP(Oporto School of Engineering). He has been teaching at FAUP since 2000 and involving in many architectural, urbanized and environmental development projects since 2001.

>>166
Emre Arolat Architects
The founding partner of Emre Arolat Architects Emre Arolat was born in 1963. Throughout his career, having participated in various conferences and seminars. He has also curated some exhibitions including the first Istanbul Design Bienniale. Has taught at several architectural schools in Turkey and abroad, including Mimar Sinan Fine Arts University, Istanbul Bilgi University, Berlage Institute for Architecture, Delft University of Technology. Became a visiting juror at Pratt Institute and Middle East Technical University. Has received many national and international awards, including Aga Khan Award for Architecture, Mies van der Rohe Award and AR Awards for Emerging Architecture.

>>34
Werner Tscholl Architekt
Werner Tscholl was born in 1955 in Latsch, South Tirol. He graduated from the University of Firenze and established the firm in 1983. Their works was exhibited at the Venice Biennale in 2002 and 2010 to 2014. He also received the Dedalo Minosse Award in 2006.

>>124
Sergio Rojo
Was born in Spain, 1975. Received a degree of Architecture from the University of Valladolid in 2001. Is an owner of Rojo Architects based in Logroño, Spain and he was also chartered at CSCAE(Superior Council of Colleges of Architects of Spain). Has been lately working on the refurbishment of old facilities, looking for new uses in order to be recovered for the real world life today.

>>66
Carlos Quintáns Eiras
Was born in Muxía, A Coruña in 1962 and graduated with degree of Architecture from the University of A Coruña in 1987. Taught at the International architecture School of Catalonia from 2002 to 2005. Has been teaching at the Department of Construction and Architecture at University of A Coruña since 1990.

>>6
Studio Gang Architects
Is a Chicago-based collective of architects, designers, and thinkers practicing internationally that was founded by MacArthur Fellow Jeanne Gang[photo above]. She uses architecture as a medium of active response to contemporary issues and their impact on human experience. Each of her projects resonates with its specific site and culture while addressing larger global themes such as urbanization, climate, and sustainability. Her work has been exhibited at the Venice Architecture Biennale, the Museum of Modern Art, the National Building Museum, and the Art Institute of Chicago. She has received an Academy Award from the American Academy and Institute of Arts and Letters in 2006.

©Sally Ryan Photography

Paula Melâneo
Is an architect based in Lisbon. Graduated from the Lisbon Technical University in 1999 and received a master of science in Multimedia-Hypermedia from the cole Supérieure de Beaux-Arts de Paris in 2003. Besides the architecture practice, she focused on her professional activity in the editorial field, writing critics and articles specialized in architecture. Since 2001, she has been part of the editorial board of the Portuguese magazine "arqa–Architecture and Art" and the editorial coordinator for the magazine since 2010. Has been a writer for several international magazines such as FRAME and AMC. Participated in the Architecture and Design Biennale EXD'1 as editor, part of the Experimentadesign team.

Nelson Mota
Graduated from the University of Coimbra in 1998 and received a master's degree in 2006 where he lectured from 2004 to 2009. Was awarded the Távora Prize in 2006 and wrote the book called A Arquitectura do Quotidiano, 2010. Is currently a researcher and guest lecturer at the TU Delft, in the Netherlands. Is a member of the editorial board of the academic journal Footprint and also one of the partners of Comoco Architects.

>>152

Guedes Cruz Arquitectos
Is an architecture firm based in Lisbon founded by Jose Guedes Cruz, graduated in Architecture from Fine-Arts School of Lisbon(ESBAL) and lectured at the ESBAL from 1979 to 1982. Cesar Marques and Marco Martinez Marinho who had been already working in Guedes Cruz Arquitectos were granted partnership. They won the Architezer A+ Award 2014, jury prize and popular choice, in the Institutional Category and several competitions. They have been also invited to numerous conferences.

Aldo Vanini
Practices in the fields of architecture and planning. Had many of his works published in various qualified international magazines. Is a member of regional and local government boards, involved in architectural and planning researches. One of his most important research interests is the conversion of abandoned mining sites in Sardini.

>>98

Erginoğlu & Çalışlar Architects
Kerem Erginoğlu[left] was born in Zonguldak in 1966 and received Graphical Presentation Achievement Award at the National Architecture Exhibition in 1992. Hasan Çalışlar[right] was born in Istanbul 1969 and has been teaching at Mimar Sinan University, Istanbul Technical University and Yıldız Technical University. He is a member of the Chamber of Architects and Istanbul Architects Association. Both graduated from the Faculty of Architecture at Mimar Sinan Fine Arts University and founded Erginoğlu & Çalışlar Architects in 1993. They received various awards at the National Architecture Awards including Building Award, AMV Young Architects and Design Award. In 2010, they also received New & Old Award from the World Architecture Festival.

>>86

Mold Architects
Iliana Kerestetzi graduated from the National Technical University of Athens in 2006 and received a Master of Science in Advanced Architectural Design(AAD) at Columbia Graduate School of Architecture, Planning and Preservation(GSAPP) NYC, USA in 2008. She established her professional practice Mold Architects in 2010 which is now based in Athens and on Serifos Island. In 2011, Mold Architects initiates the "Serifos Properties" residential development project. The practice integrates research,

>>178
Jean-Paul Viguier et Associes
Is an internationally recognized architecture, urban planning, interior, and landscape design practice lead by Jean-Paul Viguier. He graduated with Diploma in Architecture DPLG from the National School of Beaux-Arts in Paris in 1970. Received a Master in City Planning and Urban Design from the Graduate School of Design at Harvard in 1973. Over the last two decades, the practice has put its client's needs at the heart of every project to deliver innovative and elegant design solutions. The practice is experienced in designing a diverse range of public and private projects, from large scale urban planning, offices, retail, health care, education, to housing and street furniture. Based in Paris, with a satellite office in Toulouse, the practice builds award winning projects all over the world.

>>114
Mutar Arquitectos
Claudio Molina, Daniel de la Vega and Eduardo Villalobos are leading the firm. They participated in the architecture and Urbanism workshops of the Department of External Affairs at FAUM(The Faculty of Architecture at the University of Michoacana) and the projects of Urban Improvement in Peñalolén. With the sponsorship of Microsoft and its program "School of the Future", they have worked on the reconstruction of the schools such as Escuela Santa María de Peñalolén and CEEEA Centro Educacional Erasmo Escala Arriagada.

>>58
Harris Butt Architecture
Grant Harris was born 1953 and graduated from the University of Auckland, School of Architecture in 1977. He has been practicing as an architect in New Zealand since 1982 and established Harris Butt Architecture in 2007.

>>136
Eureka
Was established in 2009 by Junya Inagaki [second from the left] and Satoshi Sano [first] who has completed the Master and Doctor's Course of Architecture from Graduate School of Engineering at Waseda University. Takuo Nagai [third] and Eisuke Hori [fourth] have joined as partner of Eureka in 2009. They are also passionately committed to a variety of fields in educational activities.

>>78
Saunders Architecture
Todd Saunders, a senior architect of Saunders Architecture was born in Canada, 1969. Received a Bachelor of Design from Nova Scotia College of Art & Design in 1992 and M.Arch from the McGill University, Montreal in 1995. Established private practice, Saunders Architecture in Bergen, Norway in 1998. Has taught at the Bergen Architecture School and is currently teaching at the Cornell University in USA. Received national and international awards including Norwegian

C3, Issue 2014.7

All Rights Reserved. Authorized translation from the Korean-English language edition published by C3 Publishing Co., Seoul.

© 2014大连理工大学出版社
著作权合同登记06-2014年第125号
版权所有·侵权必究

图书在版编目(CIP)数据

休闲小筑：汉英对照 / 韩国C3出版公社编；时真妹等译. —大连：大连理工大学出版社，2014.8
（C3建筑立场系列丛书）
书名原文：C3 Vacation Stay
ISBN 978-7-5611-9452-2

Ⅰ．①休… Ⅱ．①韩… ②时… Ⅲ．①休闲娱乐－文化建筑－建筑设计－汉、英 Ⅳ．①TU242.4

中国版本图书馆CIP数据核字(2014)第182114号

出版发行：大连理工大学出版社
（地址：大连市软件园路80号　邮编：116023）
印　　刷：上海锦良印刷厂
幅面尺寸：225mm×300mm
印　　张：12
出版时间：2014年8月第1版
印刷时间：2014年8月第1次印刷
出 版 人：金英伟
统　　筹：房　磊
责任编辑：张昕焱
封面设计：王志峰
责任校对：赵姗姗

书　　号：978-7-5611-9452-2
定　　价：228.00元

发　　行：0411-84708842
传　　真：0411-84701466
E-mail：dutp@dutp.cn
URL：http://www.dutp.cn